Modeling of Sediment Transport

Edited by Davide Pasquali

Published in London, United Kingdom

Modeling of Sediment Transport
http://dx.doi.org/10.5772/intechopen.97911
Edited by Davide Pasquali

Contributors
Sylvain S. Sébastien Victor Guillou, Romuald Verjus, Marina V. Shmakova, Abhishish Chandel, Vijay Shankar, Sergio Martínez-Aranda, Pilar García-Navarro, Davide Pasquali

First published in London, United Kingdom, 2022 by IntechOpen
IntechOpen is the global imprint of INTECHOPEN LIMITED, registered in England and Wales, registration number: 11086078, 5 Princes Gate Court, London, SW7 2QJ, United Kingdom

British Library Cataloguing-in-Publication Data
A catalogue record for this book is available from the British Library

Additional hard and PDF copies can be obtained from orders@intechopen.com

Modeling of Sediment Transport
Edited by Davide Pasquali
p. cm.
Print ISBN 978-1-80355-867-7
Online ISBN 978-1-80355-868-4
eBook (PDF) ISBN 978-1-80355-869-1

Meet the editor

Davide Pasquali is a research fellow in the Department of Civil, Construction-Architectural and Environmental Engineering (DICEAA) at the University of L'Aquila, Italy. He received a master's degree (cum laude) in Civil Engineering in 2011 and a Ph.D. in Civil Engineering in 2015, both at the University of L'Aquila. His research interests include the generation and propagation of water waves, coastal hydrodynamics and morphodynamics, physical and numerical modelling of wave–structure interaction, wave energy assessment and extraction, risk analysis, and marine sediments transport.

Contents

Preface

Sediment transport is the process by which solid particles move because of natural processes (e.g., gravity or the action of a fluid in which the sediments are contained). Sometimes due to effects related to human activities (e.g., land use changes, deforestations, etc.) some processes (i.e., erosion and deposition) can be modified in their natural evolution. Because of its interdisciplinary characteristics, sediment transport has been and still is one of the most important research topics in engineering, physics, and applied mathematics.

This book includes five chapters that discuss new approaches in sediment transport estimation, how to evaluate the physical properties of porous media, and numerical approaches to sediment transport modelling.

Chapter 1 provides a general overview of sediment transport and its declination in coastal, hydraulic, and environmental engineering. It considers the natural phenomena involved in sediment transport and emphasizes possible anthropogenic causes that may induce changes in the natural physical process of sediment transport and deposition.

Chapter 2 describes the newest approaches and formulas for sediment transport evaluation. It presents a method that allows for estimating total sediment discharge by looking at the physics of the problem, that is, it is based on the balance of the force's action on the two-phase river flow. This new approach, using the power of the river flow, tries to overcome the problem of classical methods that are often developed for a specific scale or hydraulic condition.

Chapter 3 presents an experimental investigation aimed at evaluating the hydraulic conductivity (K) of porous media. The study is aimed at deepening the impact of grain size and porosity on the K value. The chapter also compares formulations in the literature and values obtained experimentally.

Moving to the numerical aspects of sediment transport modelling, Chapter 4 presents a fully resolved model based on the direct-forcing fictitious domain method. It provides a detailed analysis of the model formulation as well as proposes a validation on a simple case. Moreover, the chapter includes a series of simulation results on the sedimentation of monodisperse and polydisperse particle systems in the case of low Reynolds numbers.

Finally, Chapter 5 presents an approach to study and model geophysical surface flows involving sediment transport. The core of the approach is a combination of a 2D mathematical model and the derivation of a finite volume numerical scheme while also considering computational costs to develop efficient simulation tools. The aim is to develop a robust instrument to analyze and model complex environmental processes (i.e., sediment transport) with realistic temporal and spatial scales and maintain reasonable computational costs.

We hope this book will encourage and inspire a new generation of researchers and practitioners to advance the knowledge of sediment transport from theoretical, numerical, and experimental points of view.

Davide Pasquali
Department of Civil, Construction-Architectural,
and Environmental Engineering Department (DICEAA),
Environmental and Maritime Hydraulic Laboratory (LIAM),
University of L'Aquila,
L'Aquila, Italy

Chapter 1

Introductory Chapter: Sediment Transport

Davide Pasquali

1. Introduction

Sediment transport has been, and still is today, one of the most fascinating and challenging research topics in coastal, hydraulic, and environmental engineering. Probably first studies on sediment transport problems can be tracked back to ancient historical periods (i.e., Roman Empire, Egypt etc.) in relation to the sizing and maintenance of irrigation canals e.g. [1]. Obviously in this periods problems have been solved only by empirical trial and error method. First theoretical concepts can be referred to DuBuat (1734–1809) who was the first to talk about the concept of shear-resistance e.g. [1]. In the following years, there have been many researchers who have contributed to the topic from a theoretical, empirical or experimental point of view (e.g. Du Puit, DuBoys, Reynolds, Forchheimer, Schoklitsch Shields, kennedy, Einstein, and Bagnold) e.g. [1]. In the past, the most of the research is devoted to understand and model the physical processes in order to estimate, for example, the shear stress, the order of magnitude of the sediment transport, bed shear stress and forms.

At present, the change in land use, dams constructions, exploitation of coastal areas, deforestation and in general human activities, have expanded the pool of topics referable to sediment transport area.

Indeed, as recently underlined by [2] one of the most important component of global change can be related to the soil erosion (namely the sediment transport). Last decades have seen the decreasing trend (i.e. a decrease has been found in almost 50% of the world's rivers) in river sediment loads [2], hence the need to study possible causes investigating the mechanism behind these problems.

Therefore, the policies and strategies for future adaptations to climate change, must contain the information about the future scenarios in order to plan future actions to manage catchments and rivers e.g. [2].

Moreover, the attention for water quality and the interest, that is, beginning to be paid to natural systems (e.g. protected areas) has encouraged a holistic and multidisciplinary approach to the study of sediments transport.

There are a lot of papers, books and manuals in literature in this field. The purpose of this volume is not to replace existing literature.

On the other hand, this book is intended to collect original works and review concerning numerical and experimental investigation, theoretical works, methodological approaches, and any other technique that allows giving the actual state-of-the-art in the field of sediment transport.

2. Erosion: a natural and human-induced process

In general erosion (and sedimentation) phenomenon can be considered as a natural process referred to the particles motion e.g. [3, 4]. As a seek of example, the sediment can be eroded by the river current acting on the movable bed, or it can be moved by the raindrop impact from eroding slopes (see **Figure 1**) and conveyed downstream reaching a river or a valley, or, in the coastal areas can be moved by the actions of the currents (i.e. the longshore current) and transported along the shoreline. However, the processes inducing the solid particles detachment can be very different depending on the considered case.

The physical mechanisms by which sediment is transported downstream of a river are as suspended sediments (i.e. in the case of fine particles) or as bedload sediments (coarse particles) e.g. [3, 4] However, the great part (namely about the 70%) of sediments reaching the coastal area is suspended e.g. [4, 5].

In all these cases, the process can be considered natural, and without external forcing or changes, it will reach an equilibrium if the forcing term does not change significantly.

However, especially in the last decades, the human activities induced considerable changes in erosion/sedimentation rate. Probably, the most striking examples of the effect of human activities on sediment transport processes are the changes in land use of large watershed areas, the deployment of artificial reservoir, and the roads or railways construction. Julien P.Y. [3] highlights that the erosion rate can be (in the worst situations) 100 times greater than the natural (i.e. geological) conditions producing the modification of the runoff characteristics of internal areas also modifying the total sediment budget of the rivers (see **Figure 2**).

Moreover, dams or small deployments for electricity energy production, can result in sediment entanglement in reservoirs [6], that is, inducing the decrease of storage capacity (i.e. a decrease in energy production) e.g. [3, 4, 7]. The effect of this occurrence is a decrease of sediment input on the coastal environments, and, therefore, a generalized deficit in the sediment budget and a consequent possibility of coastal erosion. This tendency is less pronounced in coastal areas with a good dunal system that allow restoring a portion of the sediment budget. However, the urbanization growth in coastal or/and flat areas, the increase of human pressure, and the natural phenomena modified, in some case in irreversible way, the coastal dynamics e.g. [7].

Figure 1.
Examples of eroding slopes and the natural erosion/sedimentation processes – Picture courtesy of Prof. Marcello Di Risio.

Figure 2.
Example of an uncontaminated coastal areas with the presence of a coastal dune – Picture courtesy of Prof. Marcello Di Risio.

Figure 3.
Examples of "soft" (left panel) and "hard" (right panel) techniques used to act on shoreline dynamics [courtesy of M. Di Risio].

Moreover, the sediment accumulation in the upstream area can induce bed erosion, change the morphology diversity and can modify the topology of the connection between different close areas e.g. [4].

By the way often, dredging activities e.g. [8, 9] are required to collect sediment finalized to "soft" techniques (see **Figure 3**, left panel) to restore beaches or to move the sand trapped in the harbor (clean or contaminated). Another possibility to counteract coastal dynamics, is to use "hard" techniques, that is, coastal protections (see **Figure 3**, right panel), that normally induced changes in hydrodynamics and morphodynamics inducing, in some cases, modification also in the sediment transport regime.

Figure 4.
Example of river mouth. In can be observed that the mouth is protected by lateral groins and it is prone to the sedimentation [courtesy of M. Di Risio].

Therefore, it is clear that sediment transport and its fate can induce deep changes not only in the internal and in the watershed areas but can include coastal and also urban areas increasing, in some case, the vulnerability of a specific area and consequently the risk intensity.

As a seek of example, a proper maintenance of river mouths can promote regular water runoff and avoid sedimentation phenomena that can inhibit runoff. **Figure 4** shows an example of sedimentation of a river mouth.

Also, another human effect connected to the sediment transport is the water quality. Indeed, suspended sediments in water can indirectly transport both nutrients and pollution. On the other hand, it should be considered the effect of the water quality on the suspended sediments. It has to be underlined that, however, not only pollutants can be a issue for water quality, but also the presence of heavy metals and metal compounds, which in excessive amounts can be a problem, in general for the environment e.g. [10–12].

This problem can be directly related to bathing water. Indeed, the quality of the water of coastal area depends on the one hand from sea water pollution, and on the other hand from pollution and water quality coming from rivers and estuaries. Indeed, in the case of storm runoff, and in particular during extraordinary and uncontrolled storm events, there is the possibility of water discharges containing sewer overflows, high concentration of *Escherichia coli*, fecal or other bacteria that can easily modify the water quality levels e.g. [13, 14].

3. Sediment transport modeling and monitoring

In light of what has been discussed so far, it is clear the increase of attention in modeling and monitoring of sediment transport, not only by researchers but also by stakeholders and managers e.g. [15]. As underlined by [15] there are a lot of models that are able to reach this goal. However, due to the complexity of the problem, the amount of input data at the number of different spatial and temporal scales, the choice of the correct model is always a hard task.

From a general point of view the range of model types includes: (i) empirical models (ii) statistical models (iii) numerical models (iv) analytical models.

While some years ago numerical components are rare ad used only in some detailed cases, with the development of computing power at present they are a good alternative. However, there are no general indications or prescriptions in the use in the use of one or the other. It often depends on the available input data, on the dimension of the spatial scale (i.e. the dimension of the watershed), or on the physics to describe [15].

Empirical models are probably the simplest but, at the same time, the most criticized approach. The reason relies on the simplification in the assumptions, that is, the catchment is often considered homogeneous, the nonlinearities are ignored, etc. Nevertheless, they are frequently used when input parameters are limited, or in the early step of a study or a project when a fast and general overview is needed.

Statistical models, especially with the advent of artificial intelligence, are a good tool when a lot of parameters and information on the study area are available. As obvious they are powerful and allow to elaborate scenarios at different time scales. However, they often lack a physical basis and the larger the time scale (or the larger spatial scale) the greater the needed computational power.

Also, numerical models had (in particular in the last decades) a great impact on the study of sediment transport. Depending on the complexity of the problem, they can be one-, two- o three-dimensional and can be used to model sediment transport, sediment sources, river floods, eroded and deposition areas, etc. Also in this case (as for statistical models) the biggest issue is related to the computational costs. The most used, also in practical applications are probably 1D and 2D depending on the complexity of the problem. 3D models are often used to model very detailed situations requiring particular attention, that is, for environmental impact.

Finally, an important class of models is the analytical one. They are based on the solution of the equation of streamflow, hydrodynamics, advection, and diffusion, etc. It is intuitive that to model real situations, the hypothesis that has been made to write the equations can lead to an oversimplification of the problem. However, they allow modeling (accepting some simplification) of complex system with a physics-based approach. From the perspective of a modeling framework (i.e. the complexity of the model increases as the level of detail increases) approach, they can help to build a general overview of the problem, overcoming the uncertainties of empirical methods e.g. [8, 16].

An important component of sediment transport is, certain, the monitoring. In particular for long-term assessment, regular monitoring should be performed. Acquired data are fundamental in the case of numerical model validations, to calibrate empirical approaches and, in general, to check the reliability of all the described models e.g. [16].

The complexity of the study of sediment transport suggests that there is no better model to use. The choice is subjective and driven by the complexity of the case at hand.

4. Concluding remarks

Sediment transport is one of the most challenging topics in different fields (e.g. coastal, hydraulic and environmental engineering). Indeed, it appears in seas and oceans, lakes, rivers, harbors, and many other natural systems. Historically, all these aspects are related to specific research areas ranging from engineering, geology, geomorphology, biology, etc., but it is difficult to find a comprehensive overview of these topics. At present, behind the natural process inducing soil erosion, sediment transport, and deposition, human activities (i.e. dam, railways, bridges, etc.) have induced considerable changes in sediment transport rate. These changes have resulted in changes in the coastal sediment budget with consequences for shoreline dynamics, and of course, in the rivers' morphodynamics and in the water quality. In this scenario, considering also the future possible impact of climate change, the aim of this book is intended to organize in a single volume, original works, and review concerning numerical and experimental investigation, theoretical works, methodological approaches, and any other technique that allow giving the actual state-of-the-art in the field of sediment transport.

Author details

Davide Pasquali
Environmental and Maritime Hydraulic Laboratory (LIam), Civil, Construction-Architectural and Environmental Engineering (DICEAA), University of L'Aquila, L'Aquila, Italy

*Address all correspondence to: davide.pasquali@univaq.it

References

[1] Van Rijn L et al. Principles of Sediment Transport in Rivers, Estuaries and Coastal Seas. Amsterdam: Aqua Publications; 1993:**1006**

[2] Wang S, Fu B, Piao S, Lü Y, Ciais P, Feng X, et al. Reduced sediment transport in the Yellow River due to anthropogenic changes. Nature Geoscience. 2016;**9**:38-41

[3] Julien P. Erosion and Sedimentation. Cambridge University Press; 2010

[4] Serra T, Soler M, Barcelona A, Colomer J. Suspended sediment transport and deposition in sediment-replenished artificial floods in Mediterranean rivers. Journal of Hydrology. 2022;**609**:127756

[5] Vercruysse K, Grabowski R, Rickson R. Suspended sediment transport dynamics in rivers: Multi-scale drivers of temporal variation. Earth-Science Reviews. 2017;**166**:38-52

[6] Walling D, Fang D. Recent trends in the suspended sediment loads of the world's rivers. Global and Planetary Change. 2003;**39**:111-126

[7] Pasquali D, Marucci A. The effects of urban and economic development on coastal zone management. Sustainability. 2021;**13**:6071

[8] Di Risio M, Pasquali D, Lisi I, Romano A, Gabellini M, De Girolamo P. An analytical model for preliminary assessment of dredging-induced sediment plume of far-field evolution for spatial non homogeneous and time varying resuspension sources. Coastal Engineering. 2017;**127**:106-118

[9] Hayes D, Crockett T, Ward T, Averett D. Sediment resuspension during cutterhead dredging operations. Journal of Waterway, Port, Coastal, and Ocean Engineering. 2000;**126**:153-161

[10] Gao P. Understanding watershed suspended sediment transport. Progress in Physical Geography. 2008;**32**:243-263

[11] Golterman H, Sly P, Thomas R. Study of the Relationship between Water Quality and Sediment Transport: A Guide for the Collection and Interpretation of Sediment Quality Data. New York, NY: UNIPUB; 1983

[12] Agency, U. The Quality of Our Nation's Waters, A Summary of the National Water Quality Inventory: 1998 Report to Congress. (Report EPA841-S-00-001), Washington, DC. 2000

[13] Ferrarin, C., Penna, P., Penna, A., Spada, V., Ricci, F., Bilić, J., Krzelj, M., Ordulj, M., Sikoronja, M., Đuracić, I. & Others Modelling the quality of bathing waters in the Adriatic sea. Water. 13, 1525 (2021)

[14] He Y, He Y, Sen B, Li H, Li J, Zhang Y, et al. Storm runoff differentially influences the nutrient concentrations and microbial contamination at two distinct beaches in northern China. Science of the Total Environment. 2019; **663**:400-407

[15] Merritt W, Letcher R, Jakeman A. A review of erosion and sediment transport models. Environmental Modelling & Software. 2003;**18**:761-799

[16] Lisi I, Feola A, Bruschi A, Pedroncini A, Pasquali D, Di Risio M. Mathematical modeling framework of physical effects induced by sediments handling operations in marine and coastal areas. Journal of Marine Science And Engineering. 2019;7:149

Chapter 2

Sediment Transport in River Flows: New Approaches and Formulas

Marina Shmakova

Abstract

A new method for estimating the total sediment discharge, as built on balance of power acting to moving sediment particle in "water stream-bottom sediments—sediments" system, enables consideration of interrelated influence of hydraulic variables state of flow and sediment. At the same time, the basic sticking point of river hydraulics, that is, interaction of fluid and bottom, is specified not from the part of fluid boundary, but from that of bottom sediments and their properties, well studied in soil science. Setting the size of bottom sediments by means of their qualitative characteristics allows avoiding calculation errors that occur when using specific values of quantiles of bottom sediments in calculations. Consideration of the critical velocities and the phase hydraulic space of the flow allowed obtaining the equations for transporting capacity of the flow, suspended, and bed load discharges.

Keywords: river flow, suspended load, bed load, phase hydraulic space, critical velocity, transporting potential, transporting capacity of a flow, sediment discharge

1. Introduction

Solid runoff is one of the important state variables' indicators of two-phase water mass circulation in a water object. Solid runoff of a water object represents solid matter, available in river flow or moving lake water masses, and having different genesis: ground (mineral solids) or organic matter. Solid runoff of water objects can be considered in different applications: static (water turbidity), dynamic (suspended and bed load, total sediment discharge), and indirect dynamic (bed mark changes and banks transformation).

Any irregular non-stationary two-phase flow is characterized by the processes of redistribution of solid matter in the riverbed or water area of a water body. At the same time, both the processes of sedimentation and bottom erosion and the transit of sediment can be observed.

Sediment can move in river flow by means of drawing or rolling over a bed (bed load sediment), saltation, in suspended state (suspended sediment), and over flow surface due to water surface tension (flotation). And maximum possible amount of solid matter, which specified water discharge can carry over, is called flow transporting capacity. Transporting capacity of a flow defines a process of redistribution of sediment in a channel—the main factor of channel processes. If sediment discharges lesser than transporting capacity of a flow, bottom sediment is engaged into the movement, and bottom erosion occurs. When sediment discharge starts exceeding transporting capacity of a flow, sedimentation of moving sediment occurs and bed marks increase. If sediment discharge in river flow conforms to its transporting capacity, then dynamic balance is observed between suspension and settling processes.

Currently, many formulas exist for suspended and bed load sediment discharges, total sediment discharge, and flow transporting capacity calculation. At the same, the high order of arguments degree in the sediment transport formulas leads to greater calculation error. One of the most significant calculation errors for such formulas is using the bottom sediment size, which is featured by high variability in the river channel, as an argument. For objective estimation of this value, bottom sediment samples taking for the whole watercourse cross section are required that seems to be impossible in some cases. Known formulas of sediment discharge do not consider interrelated effect of flow hydraulics and transportable solid particles. Also, the issue of friction parameters setting on solid boundary of river flow, despite multi-year study of this process, still remains not enough investigated.

Sediment transport calculation algorithms, developed separately for suspended and bed load sediment in the river flow, also bring known issues in calculation accuracy. Sediment transport is caused by flow energy and the size of transportable particles, only river flow is characterized by the transport potential in accordance with which the amount (mass) of transported particles is determined. And, depending on particle size distribution of transportable material, a part of sediment uprises in water layer and migrates with water mass of flow, and a part is dragged and rolled over a bed. Moreover, the ratio of suspended and bed load kind of transport is highly variable. It depends on hydrodynamic flow pattern and solid matter income from outside (e.g., from river basin or following the result of dumping of soil in the river channel). Hydrodynamic flow pattern, in queue, is defined by channel irregularity, slope variability, and bottom sediment size, as well as water and sediment discharges of upstream.

Creating a method for calculating total sediment discharge based on the balance of forces acting in the two-phase river flow will allow for the interrelated effects of hydraulically variables of a flow state and solid matter carried by the flow. At the same time, the main stumbling block of river hydraulics is the interaction of a moving stream, and the bottom should be considered not from the side of the boundary layer of the liquid, but from the side of bottom sediment, their well-studied properties in soil science. Setting the bottom sediment size through their qualitative characterization (by categories that include wide ranges of particle coarseness variation) allows us to avoid calculation errors that arise when using specific values of bottom sediment quantiles.

2. Phase hydraulic space and critical velocities

In a river flow, it can be considered the change in the flow velocity within one water discharge (phase hydraulic space) or its average value (average flow velocity within the phase hydraulic space) for the current water discharge. In the first case, we may speak of phase hydraulic space [1]. It seems clear that the same water discharge can carry the following amount of sediment $G = [0; G_{max}]$, where G_{max} is max possible amount of sediment transportable by this water discharge per unit of time or, in other words, transporting capacity of a flow. At $G = 0$, the flow will be clarified, and its velocity will be minimal, and the depth will increase. At $G = G_{max}$, velocity increases and flow depth decreases [2, 3]. Moreover, the same water discharge (Q = const) conforms to each uttermost case. It is evident that specified water discharge at constant flow width can be defined from different combinations of depth and velocity. By plotting velocity along one axle and flow depth along another axle and indicating the points corresponding to one flow discharge value $h_i \cdot v_i = Q/B$ = const at flow width B = const in the graph, we get the function representing phase hydraulic space of a flow (system state space). Sediment discharge G_i corresponds to each depth-velocity combination ($h_i \cdot v_i = Q/B$ = const). Sediment discharge value for each combination is defined by transporting potential of a flow. That is, transporting potential of a flow represents the mass of solid matter transportable per unit of time through the cross section of a flow at constant water discharge. And this mass of solid matter, in turn, determines the ratio of velocity to depth of flow. Transporting potential of a flow is lesser than or equals to flow transporting capacity (at the same water discharge). The uttermost points of presented function are defined by process physics and conform to clarified flow (h_{max}, v_{min}) from one side and transporting capacity a flow (h_{min}, v_{max}) from another side.

Within possible flow velocity change range at specified water discharge, velocity can conform to different critical values. Critical river flow velocities traditionally are called such flow velocities at which the conditions of movement of the water flow and sediment carried by the flow change. Critical flow velocities predefine movement pattern of both liquid phase (laminar, turbulent, subcritical, and supercritical flow and so on) and solid phase of a flow. Depending on this, different types of critical velocities predefining sediment transport and bed mark change regimens are classified. For example [4]:

- non-eroding velocity (extreme velocity, whereas the basic part of bottom sediment is at rest state);
- breakaway velocity (start of mass movement of particles);
- non-silting velocity (extreme velocity, whereas the particles stay in suspended state).

Inter alia, Goncharov provides such definition for breakaway velocity: “the least average flow velocity, whereat unhampered breakaway of individual protruding grains on the bed occurs and average level of pulsation uplift forces is approximately equal to grain weight in water” [5]. Such definition enables concluding that breakaway velocity conforms to the start of particles movement both in suspended and bed load sediment forms. According to Zamarin [6], “non-silting velocity is the least average flow velocity, whereat suspended sediment, contained in water, do not leave a flow.” Obviously, the area of the clarified flow will be characterized by the lower limit of non-eroding velocity, while the area of maximum suspended load on the flow will be characterized by the non-silting velocity. Moreover, it is implied that the start of particles movement in river flow, predominantly, by saltation and in suspended form, conforms to velocity, being within the range between non-eroding and breakaway velocities. The latter can be explained by the fact that when the transporting potential of the river flow falls, particles of the largest size are the first to be deposited on the bottom during the movement of multi-factional sediment. Upper layer of deposited sediment will be represented by fine-grain fractions. Then, if transporting potential increases, at first, the upper layer of bottom sediment, represented by lesser size particles, starts movement. In support of the latter, let us consider the expressions, defined for critical velocities of movement start by drawing or rolling over a bed, $v_{cr\ bed}$, and suspensing, $v_{cr\ suspend}$ [7]:

$$v_{cr\,bed} = 5.75\lg\left(2\frac{h}{d_{50}}\right)\sqrt{\theta_{cr\,bed}\left(\frac{\rho_s}{\rho_w}-1\right)gd_{50}}, \tag{1}$$

$$v_{cr\,suspend} = 5.75\lg\left(\frac{12h}{6d_{50}}\right)\sqrt{\theta_{cr\,suspend}\left(\frac{\rho_s}{\rho_w}-1\right)gd_{50}}, \tag{2}$$

$$\theta_{cr\,bed} = \frac{0.3}{1+1.2D^*} + 0.055(1-\exp[-0.02D^*]), \tag{3}$$

$$\theta_{cr\,suspend} = \frac{0.3}{1+D^*} + 0.1(1-\exp[-0.05D^*]), \tag{4}$$

$$D^* = d_{50}\left[g\frac{\frac{\rho_s}{\rho_w}-1}{v^2}\right]^{\frac{1}{3}}, \tag{5}$$

where d_{50}—particle diameter with probability 50%, m; g—acceleration of gravity, m/s^2; ν—kinematic viscosity coefficient, m^2/s; ρ_s and ρ_w—densities of soil and water, respectively, kg/m^3; h—average flow depth, m.

From these ratios, it follows that the movement of suspended particles, whose particle size is less than 30% of the size of the particles moving by drawing, is determined by a smaller value of the critical velocity. In a context of big grain size variety of mineral particles, available in river channel, this means that the start of sediment movement falls to suspended form.

The phase hydraulic space is characterized by the morphometry of the channel and the nature of the underlying surface, and the transporting potential of the flow is determined in accordance with the amount of solid matter entering the flow. Type of the function approximating the phase hydraulic space is defined by cross section form, and the function itself represents the velocity-to-depth ratio for constant water discharge at studied cross section. **Figure 1** provides an example of phase hydraulic space for a channel with rectangular shape. In this case, change range for flow velocities and depth is defined by water discharge. It seems to be clear that water discharge predefines opportunity to reach this or that critical velocity. Thus, within the limits of one water discharge, a different ratio of hydraulic variables of a flow state can be established. This ratio is strongly predefined by solid matter ingress from catchment.

The flow velocities achieved within the phase hydraulic space for a fixed water discharge were considered above. Let us now stop at the average velocity, which corresponds to a given water discharge. If sediment in the river flow is formed solely by means of channel deformation, then we speak of channel-forming water flows. In this case, the average rates of the beginning and end of the process of channel deformation correspond to certain critical values. Then, the ratio of critical velocities and sediment transport pattern comes to the following schematic (**Figure 2**) (the view of this dependence is conditional).

At the same time, both for the phase hydraulic space (within a fixed water discharge) and for the average velocity for a given water discharge, there is a ratio of critical flow rates and the nature of transport of multi-fractional sediment:

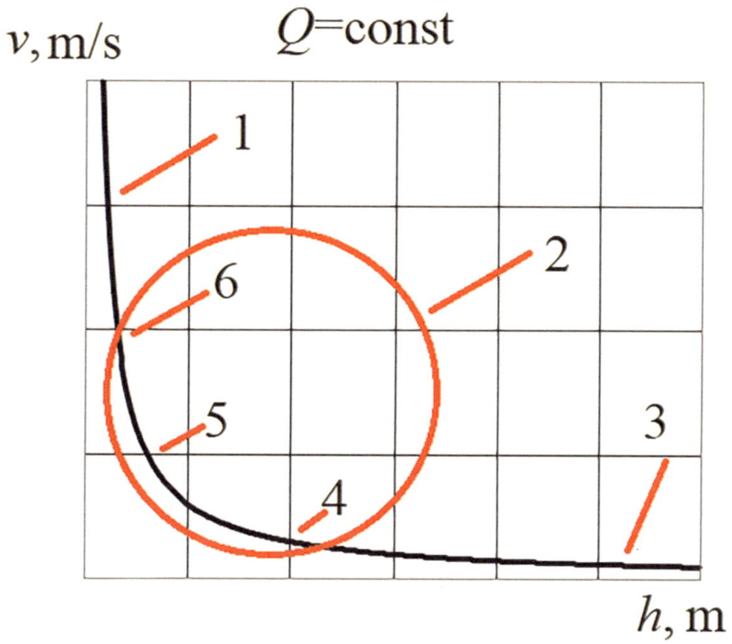

Figure 1.
Phase hydraulic space of a flow in the channel cross section for Q = const: 1—the area of the maximum suspended load on the flow; 2—the area of physically possible values; 3—clarified flow area; 4—non-eroding velocity (start of the movement of suspended sediment); 5—breakaway velocity (movement of suspended and bed load sediment); 6—non-silting velocity.

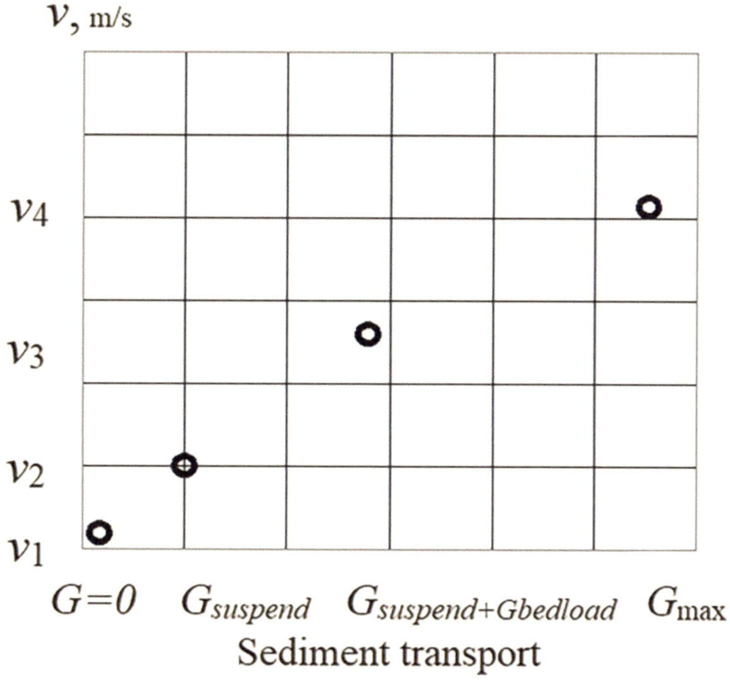

Figure 2.
Correlation of critical velocities and type of sediment transport.

- v_1—velocity lesser than minimal non-eroding velocity (no sediment movement, $G = 0$);
- v_2—non-eroding velocity (start of suspended sediment movement, $G_{suspend}$);
- v_3—breakaway velocity (start of bed load and suspended sediment movement, $G_{suspend\ +\ bedload}$);
- v_4—non-silting velocity (transporting capacity of a flow, $G_{\max}$).

3. Review of sediment discharge formulas

In the early days of research into the movement of solid material in river flow, a separation of total sediment into a suspended and a bed load sediment part was adopted. Such posing of the question was justified by measurement base opportunities and useful to form general representation on regularities of involvement into movement and solid matter transfer in river flow. However, amid modern views on process physics, vast experience of full-scale and laboratory experiment, instrument opportunities, and all-round interdisciplinary integration, such vision of the issue seems slightly limited.

3.1 Suspended sediment discharge formulas

Development of suspended sediment discharge formulas, unfortunately, failed to find enough place in hydraulic calculations practice. At the same time in investigation of suspended sediment, the big attention was paid to turbidity distribution along vertical and along river length. Furthermore, vast investigations with another averaging scale were conducted on generalized materials analysis of spatiotemporal turbidity distribution (Karaushev [8], Shamov [9]). And as formulas for the consumption of suspended sediment in the standards and applied works, formulas for calculating the transporting capacity of a flow are proposed. Admixture propagation equation based on diffusion theory of sediment movement (Taylor [10], Schmidt [11], Makkaveev [12], Karaushev [8]) is represented in educational and scientific literature for suspended sediment discharge calculation. However, neither in the first case (formulas of transporting capacity of a flow) nor in the second case (admixture propagation equation), the task of suspended matter concentration estimation in a flow is not solved finally. It is evident that suspension-bearing load of river flow not always conforms to its transporting capacity. And admixture propagation equation, being an elementary continuity equation at known suspended matter concentration in a flow (boundary condition specification), leaves the question of estimating this concentration open.

In the last century, some suspended sediment discharge formulas were also based on knowledge on suspended matter concentration in a near-bottom layer and came down to integral calculation of suspended matter concentration distribution epure. In 1937, Rouse [13] provided theoretic equation for vertical distribution of suspended particles in turbulent flow. Suspended matter concentration formulas by Karaushev [8], Van Rijn [14], Sedaei et al. [15], Bagnold [16], Karasev [4], etc., are used in hydraulic calculations practice.

3.2 Bed load sediment discharge formulas

As is commonly known, assessment of bed load sediment discharge for natural water objects is one of the most complicated hydraulic tasks. Disconcertingly, lately activity in developing new approaches to solve this issue is not enough. Moreover, the main encumbrance lays in the absence of reliable verification for proposed calculation formulas according to field studies data. Bed load sediment discharge formulas can be focused both on the movement of separate solid particles directly or ridge form of sediment movement. Formulas describing ridge form of sediment movement consider geometric parameters of ridges, their length, height, etc., and can be used for rivers with sand bed. Such formulas are supported with relatively true data of sediment discharge observation that enables optimization of both formula's structure and its parameters. Bed load sediment in rivers with gravel bed represents the biggest complexity in sediment discharge measurements. This, accordingly, hampers probation of bed load sediment discharge formulas and optimization of structure and parameters of such formulas.

Some researchers classify the following groups of bed load sediment discharge formulas:

1. Dependence of sediment discharge on hydraulic flow characteristics (Shamov [9], Levi [17], Goncharov [5], Grishanin [18], Egiazarov [19], Van Rijn [20]);

2. Relation of sediment discharge to water content (water discharge) of a river (Meyer-Peter et al. [21], Meyer-Peter and Müller [22], Schoklitsch [23, 24], Gilbert [25]);

3. Relation of the sediment discharge to the attractive force of the flow (Egiazarov [19], Bagnold [16]);

4. Formulas, where stochastic nature of sediment movement is expressed (Einstein [26, 27], Velikanov [28], Shen and Hung [29]).

Given enough great quantity of bed load sediment discharge formulas, but total sediment discharge formulas are not as common. However, at empirical nature of bed load sediment discharge formulas, the formulas for total sediment discharge are often more physically based. Total sediment discharge is a function of hydraulic flow parameters, such as average flow velocity, depth, water discharge, slope, size, hydraulic size, and density of the particles, as well as shear stress on solid boundary of a flow. Some formulas are developed based on dimensional analysis and almost all of them based on main concept of shear force of a flow.

The formula by Yang and Lim is defined using dimensional analysis for rivers with sand bed [30]. The formula by Ackers and White is also derived from dimensional analysis [31, 32]. The transport of fine-dispersed material is associated with a shear velocity, and the transport of larger particles is associated with an average flow velocity. Karim and Kennedy also obtained a formula for the total sediment through the theory of dimensions, making the total sediment flow dependent on the average and dynamic flow velocity, hydraulic size, and average particle size [33, 34]. Yang hypothesized that the determining factor in the concentration of sediment in alluvial channels is the specific power of flow, which can be defined as the calculated per unit time dissipation of potential energy per unit weight of water [35, 36]. The formula by

Engelund and Hansen, defined in the middle of last century [37], is based on Bagnold's flow power concept and theory of similarity. The formula by Molinas and Wu is based on shearing force of a flow [38]. In this formula, the Darcy-Weisbach equation is solved together with an expression for the frictional force, which gives the relationship between the total sediment concentration and the resulting flow force. The formula by Bagnold [16, 39] is based on energy balance concept, where flow power predefines energy for sediment transport. The formula of total sediment discharge developed by Karasev is based on two dependences for suspended and bed load sediment discharges [4]. The commonality of the mechanisms of movement consists of a single process of interaction between a liquid and a solid medium characterized by turbidity of ascent [4].

3.3 Formulas of transporting capacity of a flow

The values of empirical coefficients in the formulae of sediment discharge can be determined by minimizing the deviations between the results of calculations and observational data. But when deriving the formulas of flow transporting capacity, it's hard to focus on observation data for definite river, as limit flow saturation with sediment is not achieved on all rivers and not for all water content periods.

One of the determining factors for the involvement of sediment particles in the flow is the turbulence regime of the river—velocity pulsations have a suspending and supporting effect on the particles in the flow. But it is known that the presence of a solid substance in the flow significantly reduces the pulsations of velocities, the flow becomes relatively orderly. Whereas all other things being equal, the clarified stream, having a large erosion capacity of the channel, has a more turbulent mode of movement.

In can therefore be concluded that the degree of flow saturation with sediment has nonlinear dependence from average flow velocity and, among others, depends on flow movement pattern.

Therefore, when derivation of the formulas of flow transporting capacity, not only dependence on the amount of transported matter from hydraulic variables of flow state shall be considered, but the factors predefining limit fluid saturation with suspensions. Such factors can include suspending capacity of a flow, Froude and Reynolds criteria, as well as presence of small fractions in a flow. It is known that high content of finest particles increases water viscosity and, therefore, impacts onto flow capacity to transport coarser fractions. Accordingly, "the limit of flow saturation with sediment depends on both flow hydraulics and transported sediment content" [8]. The formulas by Zamarin [6], Bagnold [39], Karaushev [8], and others are known from assessment practice of flow transporting capacity $G_{\max}$.

4. New approaches to sediment transport assessment

4.1 Formula of total sediment discharge

Main equation of mathematic model for water and solid matter movement in river flow is based on balance of forces acting to moving sediment particle in "water flow—bottom sediment—sediment" system. The forces acting on the particle side are counteracted by the forces acting on the flow side:

$$\vec{F}_{flow} + \vec{F}_s = 0. \tag{6}$$

Total force balance equation in "water flow—bottom sediment—sediment" system has the following view:

$$\vec{F}_{grav} + \vec{F}_{inert} + \vec{F}_A + \vec{F}_{grav\,s} + \vec{F}_{resist} + \vec{F}_{inert\,s} = 0, \tag{7}$$

where $\vec{F}_{grav}$—shear projection of the flow gravity; $\vec{F}_{inert}$—the force of inertia of the moving volume of water enclosed between the sections; $\vec{F}_A$—Archimedes force; $\vec{F}_{grav\,s}$—a retaining projection of gravity acting on a sediment particle moving in a flow; $\vec{F}_{resist}$—bottom resistance force; $\vec{F}_{inert\,s}$—the force of inertia of a particle moving in a flow.

Bottom resistance force F_{resist} is written by analogy with the well-known formula of the linear relationship between soil shear resistance and normal load [40]:

$$F_{resist} = Rf + cS, \tag{8}$$

where f—internal friction coefficient, dimensionless; c—adhesion of soil particles during shear kg/(m·s^2) (for disconnected soil $c = 0$); S—the area of force application, m^2; in this case, the load R represents the system of loads (F_{flow}) acting to soils particles from flow side:

$$R = F_{flow} = \vec{F}_{grav} + \vec{F}_{inert} + \vec{F}_A \tag{9}$$

Therefore, main equation of water and solid matter movement has the following view:

$$(1-f)\left(mg\left[I - \frac{\partial h}{\partial x}\right] - m\frac{dv}{dt}\right) - N_{act}m_s\frac{dv_s}{dt} + N_{act}m_s g - cS = 0, \tag{10}$$

$$\frac{\partial h}{\partial t} + h\frac{\partial v}{\partial x} + v\frac{\partial h}{\partial x} = 0, \tag{11}$$

$$v_s = \sqrt{v^2 + \omega^2}, \tag{12}$$

$$\frac{dE_{flow}}{dt} - \frac{dE_s}{dt} = 0, \tag{13}$$

where m—the mass of the volume of water enclosed between the two cross sections, kg; g—acceleration of gravity, m/s^2; I—bottom slope, dimensionless; v—flow velocity, m/s; h—flow depth, m; x—longitudinal coordinate, m; S—force application area, m^2; m_s—particle mass, kg; v_s—particle velocity, m/s; N_{act}—the number of moving particles in the flow; ω—hydraulic particle size, m/s; E_{flow}—kinetic energy of the flow, kg·m^2/s^2; E_s—kinetic energy of moving particles, kg·m^2/s^2; f—coefficient of internal friction, dimensionless; c—the parameter of adhesion of soil particles during shear, kg/(m·s^2).

Water and solid matter movement equation (Eq. (10)) is closed by flow continuity equations (Eq. (11)), equations for particle velocity (Eq. (12)), and the equation of flow and particles kinetic power balance (Eq. (13)).

For conditions for uniform steady movement after some transformations of the equation (Eq. (10)), we can acquire that sediment discharge G, that is, the mass of solid matter passing through flow cross section per unit of time (kg/s), shall be equal to:

$$G = \frac{N_{act} m_s}{\Delta t} = \frac{\left[\frac{cBv \cdot \Delta t}{g} - (1-f)mI\right]}{\Delta t}, \tag{14}$$

where B—the width of the stream, m. Or, if we write that $m = \rho_w \cdot h \cdot B \cdot v \cdot \Delta t$ and water discharge $Q = v \cdot h \cdot B$, (m^3/s) we get:

$$G = Q\left[\frac{c}{hg} - (1-f)I\rho_w\right], \tag{15}$$

Thus, upon reductions and transformations, two basic groups remain in the formula: gravitational component ($\rho_w \cdot m \cdot I$) and soil shear strength or friction force ($c/(h \cdot g) + \rho_w f \cdot m \cdot I$).

It worth noting that formula (Eq. (15)) calculates the mass of solid matter in water and this mass shall be brought to real mass of solid matter:

$$G' = G\frac{\rho_s}{\rho_s - \rho_w}. \tag{16}$$

Therefore, the formula (Eq. (15)) takes the following view:

$$G = \frac{\rho_s}{\rho_s - \rho_w} Q\left[\frac{c}{hg} - (1-f)I\rho_w\right]. \tag{17}$$

Parameters f and c of the formula (Eq. (17)) depend on water content phase of a river and size of bottom sediment. Dependences on different water content periods can be used to define the values of parameter f (for the group of the studied rivers):

$$f_1 = -0.1293D + 1.7143, \tag{18}$$

$$f_2 = -0.0477D + 1.2937, \tag{19}$$

$$f_3 = -0.0114D + 1.0556, \tag{20}$$

or in common view

$$f_i = a_i D + b_i, \tag{21}$$

where i—the water content index (1—maximum, 2—average, 3—minimum water content); D—a qualitative sign of the size of bottom sediment: 1—loam; 2—sand; 3—sand-pebbles; 4—gravel; 5—pebbles.

Following the optimization, the values of parameter c for connected soil (clay loams) for the studied rivers are equal to 0.385, 0.505, and 1.55 kg/(m·s^2) in average for the periods of low-, medium-, and high-water content, respectively. For loose soils for the most of studied rivers within medium- and high-water content obtained values of parameter c following the optimization were equal to zero.

The given values of the parameters f and c are preliminary and may be useful for unexplored rivers. For rivers where hydraulic variables of flow state were measured, the values of these parameters are estimated by minimizing the deviation of the calculated and observed values of the sediment discharges.

4.2 Formula of flow transporting capacity

Let us consider deriving the formula of flow transporting capacity. Based on phase hydraulic space concept, we assume that flow transporting capacity $G_{\max}$ is equal to sediment discharge at max flow velocity (and minimal depth) for specified water discharge, that is,

$$G_{\max} = \frac{\rho_s}{\rho_s - \rho_w} Q \left[\frac{c}{h_{\min} g} - (1 - f)\rho_w I \right], \tag{22}$$

where $h_{\min}$—minimum possible depth at fixed water discharge, slope and bottom sediment size, m.

Minimum possible depth is predefined by channel's morphometry and water discharge. A decrease in $h_{\min}$ is already practically impossible, since within the value of this depth there will be an intensive suspend of solid matter to the bottom.

Thus, the calculation of the transporting capacity of the flow is preceded by an estimate of the values of $h_{\min}$ or $v_{\max}$.

To estimate the boundary velocity $v_{\max}$ (so-called silting velocity), at which the suspend in the flow either precipitates or the qualitative state of the flow changes—the flow becomes viscous, we can use, for example, the expression [41]:

$$v_{\max}^2 = \frac{\rho_s}{\rho_s - \rho_w} \frac{1}{2} g h. \tag{23}$$

According to the formula (Eq. (23)), minimal flow depth will be:

$$h_{\min} = \sqrt[3]{2 \frac{\rho_s - \rho_w}{\rho_s} \frac{Q^2}{B^2 g}}. \tag{24}$$

Then the expression for transporting capacity of a flow (Eq. (22)), considering the formula (Eq. (24)) will be as follows:

$$G_{\max} = \frac{\rho_s}{\rho_s - \rho_w} Q \left[\frac{c}{g \sqrt[3]{2 \frac{\rho_s - \rho_w}{\rho_s} \frac{Q^2}{B^2 g}}} - (1 - f)\rho_w I \right]. \tag{25}$$

Therefore, defined analytical formula for transporting capacity of a flow (Eq. (25)) is based on balance of forces acting in the system "water flow—bottom sediment—sediment" [1], the formula of soil shear resistance [40], and the formula of the boundary velocity of particle deposition in the water flow [41]. Probation of formula (Eq. (25)) and comparative analysis of some formulas of flow transporting capacity are provided in [1].

4.3 Suspended and bed load sediment discharge formulas

Consider now the derivation of formulas for suspended and bed load sediment. The start of sediment suspension is due to increase in average flow velocity against non-eroding velocity. The formula of non-eroding velocity v_{cr}, defined based on multiple studies, is as follows [42]:

$$v_{cr} = 1.15\sqrt{g}(h_{cr}d)^{0.25}. \tag{26}$$

By expressing the depth from (Eq. (26)), we will get:

$$h_{cr} = \frac{v_{cr}^4}{1.15^4 g^2 d}. \tag{27}$$

By substituting the defined expression (Eq. (27)) to total sediment discharge formula (Eq. (17)) we'll get suspended sediment discharge formula, $G_{suspend}$

$$G_{suspend} = \frac{\rho_s}{\rho_s - \rho_w} Q\left(1.15^4 \frac{cgd}{v^4} - (1-f)I\rho_w\right). \tag{28}$$

Since total sediment discharge represents the sum of suspended and bed load sediment discharges, we'll get the expression for bed load sediment discharge, $G_{bedload}$

$$G_{bedload} = G - G_{suspend}, \tag{29}$$

$$G_{bedload} = \frac{\rho_s}{\rho_s - \rho_w} Q\left(\frac{c}{gh} - (1-f)I\rho_w\right) - \frac{\rho_s}{\rho_s - \rho_w} Q\left(1.15^4 \frac{cgd}{v^4} - (1-f)I\rho_w\right), \tag{30}$$

$$G_{bedload} = \frac{\rho_s}{\rho_s - \rho_w} Qc\left(\frac{1}{gh} - 1.15^4 \frac{gd}{v^4}\right). \tag{31}$$

This section thus presents the derivation of four formulas: total sediment discharge formula (Eq. (17)), flow transport capacity formula (Eq. (25)), suspended sediment discharge (Eq. (28)), and bed load sediment discharge (Eq. (31)) formulas. All derived formulas are based on the equation of water and solids motion (Eq. (10)), the concept of phase hydraulic space, and the relationships describing the critical flow states.

5. Calculation results for derived formulas

Consider the results of approbation of the new sediment discharge formulas. Observation data on 15 hydrometric stations, located on American rivers in the states Alaska, Idaho, Colorado, Washington, and Wisconsin, were used as calculation material. Observations on these rivers and creeks were conducted in 70s–80s years of the last century, and investigation results are represented in the report "Measured total sediment loads (suspended and bed load sediment) for 93 United States streams." This report is published on official website of Geological Survey at the US Home Department and is in public domain [43]. The report represents the data on suspended and bed load sediment measured nearly simultaneously. Besides this, the report provides hydraulic variables of flow state and и grain-size analysis of sediment and bottom sediment. "The data, most of which were not published earlier, were measured by means different individuals and entities ... Despite known sampling issues, the data are, probably, the best of those available for the moment" [43] (for 1989).

The report provides observation results in 93 rivers and creeks; however, the most comprehensive data required for calculation are represented for 15 rivers only. Totally, 252 measurement data for medium water content period were used in calculations. The range of main hydraulic characteristics of the studied rivers, wherein calculations were conducted, is represented in **Table 1**.

No	Hydrometric section	I, non-dim.	Q, m^3/s	v, m/s	h, m	B, m	$G_{bedload}$, kg/s	$G_{suspend}$, kg/s	G, kg/s
1	Susitna River near Talkeetna, Alaska	0.00146	612–1160	1.8–2.7	1.7–2.3	183–202	197–849	2.26–10.4	199–859
2	Chulitua River below Canyon near Talkeetna, Alaska	0.00131	564–946	2.0–2.5	2.5–3.1	108–123	462–1690	26.9–145	488–1835
3	Tanana River at Fairbanks, Alaska	0.000467	1010–2020	1.3–1.9	2–2.9	296–469	2410–10,030	16.3–89.4	2426–10,119
4	Snake River near Anatone, Wash	0.001033	1990–3770	2.4–3.3	4.7–5.8	181–197	61.6–1270	1.07–58.7	62.67–1328
5	Toutle River at Tower Road near Silver Lake, Wash.	0.00311	112–248	1.5–3.1	0.77–1.5	61–70	538–5820	33–232	682–5901
6	Fork Toutle River near Kid Valley, Wash.	0.0037	110–185	2.4–2.8	0.85–1.1	56–59	1590–4980	110–338	1883–5090
7	Clearwater River at Spalding, Idaho	0.000312	847–1810	1.4–2.5	4.3–5.1	135–143	15.8–124	0.578–6.54	16.38–130.54
8	Yampa River at Deerloge Park, Colo	0.000673	108–447	0.81–1.3	1.5–3.9	90–93	113–998	3.6–13.2	122.3–1008
9	Wisconsin River at Muscods, Wis.	0.000311	114–714	0.49–0.88	0.71–2.6	278–310	1.42–20	1.76–23.3	3.18–43.3
10	Black River near Galesville, Wis.	0.000221	20.1–80.7	0.44–0.54	0.55–1.4	72–122	0.422–5.5	1.28–4.09	1.702–9.59
11	Chippewa River at Durand, Wis.	0.000326	132–884	0.77–1.1	1.3–3.2	215–244	2.78–64.5	5.52–23.3	8.3–87.8
12	Chippewa River near Pepin, Wis.	0.000309	118–391	0.57–0.86	10.76–1.8	229–274	2.24–45.7	2.89–14.7	5.13–60.4
13	North Fork of Lick Creek near Yellow Pine, Idaho	0.00666	1.28–4.25	0.52–0.95	0.32–0.51	7.3–8.8	0.00294–0.0978	0.0005–0.0292	0.00344–0.127
14	South Fork of Salmon River near Cascade, Idaho	0.00695	22–77.5	0.62–1.3	1.1–1.7	31.5–34.5	0.132–4.11	0.00632–6.42	0.138–10.53
15	Chippewa River near Caryville, Wis.	0.000213	117–779	0.45–1.1	1.4–2.8	185–247	0.936–16.4	0–13.5	0.936–29.9

Table 1.
The main hydraulic characteristics of the studied rivers.

On average, the deviations between the observed and calculated values for the medium water content period have been: according to the formula of total sediment discharge (Eq. (17))—41%; according to the formula of suspended sediment discharge (Eq. (28))—48%; according to the formula of bed load sediment discharge (Eq. (31))—46%. The results obtained are quite acceptable and confirm the operability of the above formulas.

It should be noted that calculation results for the formulas derived by other authors, as provided in this work, were published earlier many times, including [1]. At the same, the best results are shown by the formulas [1]:

- Engelund and Hansen's [37]:

$$G = \rho_s \cdot B \frac{0.05|v|^5}{(\rho_s/\rho_w - 1)\sqrt{g}d_{50}C^3}, \quad (32)$$

- Karim and Kennedy's [34]:

$$G = B \cdot k \left[\frac{v}{\sqrt{g(\rho_s/\rho_w - 1)d_{50}}} \right]^{2.97} \left(\frac{u^*}{\omega} \right)^{1.47} \sqrt{g(\rho_s/\rho_w - 1)d_{50}^3}, \quad (33)$$

- Bagnold's [39]:

$$G = \frac{|v|^3}{C^2} \left(0.24 + 0.01 \frac{u^* C}{\omega g^{1/2}} \right) \rho_w, \quad (34)$$

where k—coefficient of proportionality equal to 0.00139; τ—shear stress at the bottom, kg/(m·s^2); C—Shezi's coefficient, m$^{0.5}$/s.

Figure 3 shows the relationship between the total sediment discharges observed and the calculated according to the above formulas for the study rivers. As can be seen from the graphs, the points of the observed and formula-calculated sediment discharges are almost bisecting each other. Small values of sediment load discharge are better calculated using the Karim-Kennedy's (Eq. (33)) and Engelund-Hansen's (Eq. (32)) formulas, but at the same time, the points of observed and calculated sediment discharge using the Engelund and Hansen's formula (Eq. (32)) have a greater scatter and a systematic bias toward underestimation of calculated sediment

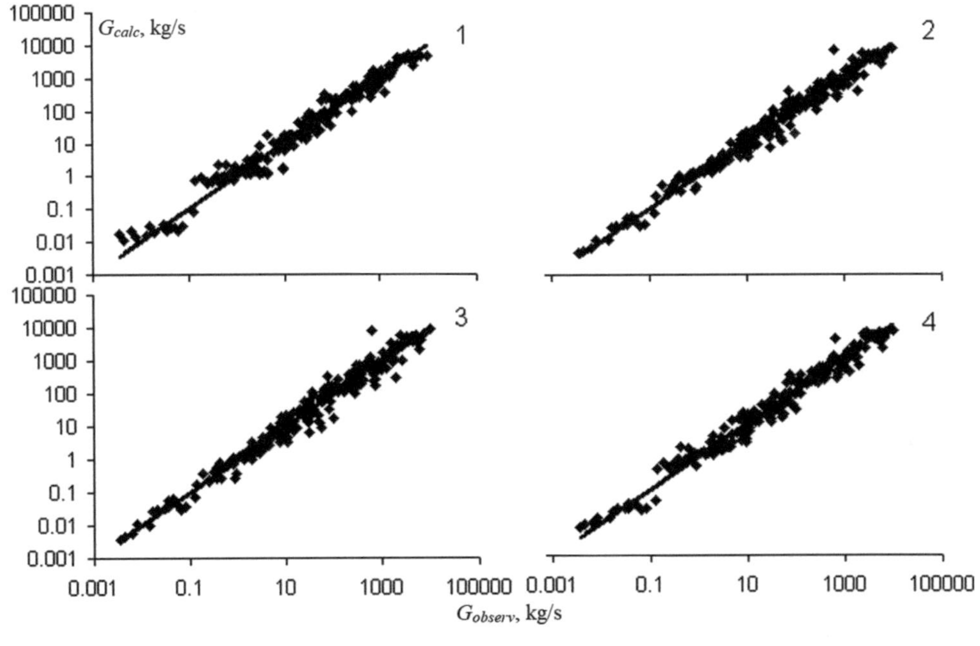

Figure 3.
Observed G_{observ} and calculated G_{calc} total sediment discharge values using Shmakova's formula (Eq. (17))—1, Karim-Kennedy's formula (Eq. (33))—2, Engelund-Hansen's formula (Eq. (32))—3, and Bagnold's formula—(Eq. (34))—4.

discharge for large values of sediment load. At the same time, Shmakova's formula (Eq. (17)) and Bagnold's formula (Eq. (34)) showed the best result in the area of large values and the worst in the area of small values.

Now consider the results of the validation of formula (Eq. (25)) and conduct a comparative analysis of some formulas for the transport capacity of the flow [1]:

- Zamarin's formula for hydraulic particle size $0.002 < \omega < 0.008$ m/s [6]:

$$G_{\max} = Q0.022\left[\frac{v}{\omega}\right]^{\frac{3}{2}}\sqrt{hI}; \tag{35}$$

- Goncharov's formula [5] for $u^*/\omega < 2.5$:

$$G_{\max} = \rho_s Q \frac{1+\phi}{800}\frac{d}{h}\frac{v_{cr}}{v}\left(\frac{v^3}{v_{cr}^3} - 1\right)\left(\frac{v}{v_{cr}} - 1\right), \tag{36}$$

$$\phi = \sqrt{\frac{\rho_s - \rho_w}{0.9\rho_w}\frac{gd}{\omega^2}}, \tag{37}$$

$$v_{cr} = 0.96\sqrt{gd^{0.4}(d + 0.0014)^{0.6}}\left(\frac{h}{d}\right)^{0.2}; \tag{38}$$

- Bagnold's formula [39]:

$$G_{\max} = Q\rho_s \frac{\rho_w}{\rho_s - \rho_w}\frac{C_f v^2}{gh}\left(\frac{0.13}{f - I} + \frac{0.01}{\omega/_v - I}\right), \tag{39}$$

$$C_f = \left(\frac{k}{\ln\left(12h/_\Delta\right)}\right)^2, \tag{40}$$

$$\Delta = \begin{cases} 3d_{90} & \text{for } \dfrac{\rho_w}{\rho_s - \rho_w}\dfrac{u_*^2}{gd} < 1 \\ 3\dfrac{\rho_w}{\rho_s - \rho_w}\dfrac{u_*^2}{gd}d_{90} & \text{for } \dfrac{\rho_w}{\rho_s - \rho_w}\dfrac{u_*^2}{gd} \geq 1 \end{cases}, \tag{41}$$

where v_{cr}—non-eroding velocity, m/s; Q—water discharge, m^3/s; I—bottom slope, dimensionless; v—average flow velocity, m/s; h—average flow depth, m; f—coefficient of internal friction, dimensionless; ω—hydraulic particle size, m/s; C_f—coefficient of friction; k—Carman constant, equal to 0.41; Δ—effective roughness height, m; d, d_{90} и d_{95}—particle diameter with probability 50, 90 и 95%, respectively, m; ρ_s и ρ_w—densities of soil and water, respectively, kg/m^3; g—acceleration of gravity, m/s^2; u^*—dynamic velocity, m/s.

The main criterion for the quality of the calculations will be the condition that the observed values of the total sediment discharge do not exceed the calculated values. This somewhat tentative and rather qualitative assessment of the correctness of the formulas can be explained by the fact that there is no information for the study rivers on how much suspended load of the flow reaches its maximum possible values. Or, in other words, it is not clear whether the measured sediment discharge G_{observ} corresponds to the flow transport capacity $G_{\max}$. A similar qualitative analysis is also allowed, for example, in [6]. This qualitative assessment is supplemented by another requirement—the calculated values must deviate reasonably from the measured values. That is, the calculated maximum sediment discharges $G_{\max\ calc}$ in its value should correspond to the hydraulic conditions of the flow. In general terms, the conditions of conformity set out can be written as:

$$\begin{cases} G_{observ} \le G_{\max calc} \le k \cdot G_{observ} \\ k \in \left[1, \dfrac{G_{\max}}{G_{observ}}\right] \end{cases}, \quad (42)$$

where k—coefficient determining the degree to which the canal is filled with moving sediment (based on observations), dimensionless.

For formula (Eq. (25)), the friction parameters were assumed to be equal to those optimized for the high and medium water periods in the calculations using the analytical sediment flow formula (Eq. (17)).

Table 2 shows the relative number of cases of non-exceeding δ, % of the observed values of sediment discharge by calculated ones, average relative deviations σ_{total}, % (between the calculated values of the flow transport capacity and the observed total sediment discharge) and $\sigma_{\min}$, % (only between the calculated values of the flow transport capacity, not exceeding the observed values, and the observed total sediment discharge). The latter indicator illustrates the degree of deviation toward a clearly erroneous calculation—as the calculation of $G_{\max}$ assumes that condition (Eq. (42)) is fulfilled.

Results of the calculations according to the four formulas are shown on the **Figure 4**. Degree of qualitative correspondence between the calculated values of total sediment discharge and the observed values is demonstrated by the excess of calculated points over the bisecting lines.

The results of the calculations above illustrate, in summary form, that in general (55–76%) the calculated sediment discharges are higher than the observed values. However, the values calculated using Shmakova's formula (Eq. (25)) are the most consistent with the order of magnitude of the observed sediment discharges, with a deviation σ_{total} of about 87%. For the same formula, the most adequate values $\sigma_{\min}$ (66%) were also obtained.

The calculations were based on a quantitative assessment of the quality of the calculation formulas by comparing the calculated values from the above formulas and the observed values of the transport capacity of the flow. For the study watercourses at the observed average flow depth, the silting velocities $v_{\max}$ were calculated using formula (Eq. (23)). These velocities were compared with the observed velocities v, to which the average depths correspond. If the velocities v and $v_{\max}$ are approximately equal, it can be assumed that the measured sediment flow rate is the transport capacity of the flow $G_{\max\ observ}$. The detected $G_{\max\ observ}$ values were compared with the $G_{\max}$ values calculated from the flow capacity formulas above.

Formula	δ, %	σ_{total}, %	$\sigma_{\min}$, %
Zamarin's (Eq. (35))	45	555	671
Goncharov's (Eq. (36))	38	819	882
Bagnold's (Eq. (39))	24	849	434
Shmakova's (Eq. (25))	33	87	66

Table 2.
Results of calculations using the flow capacity formulas.

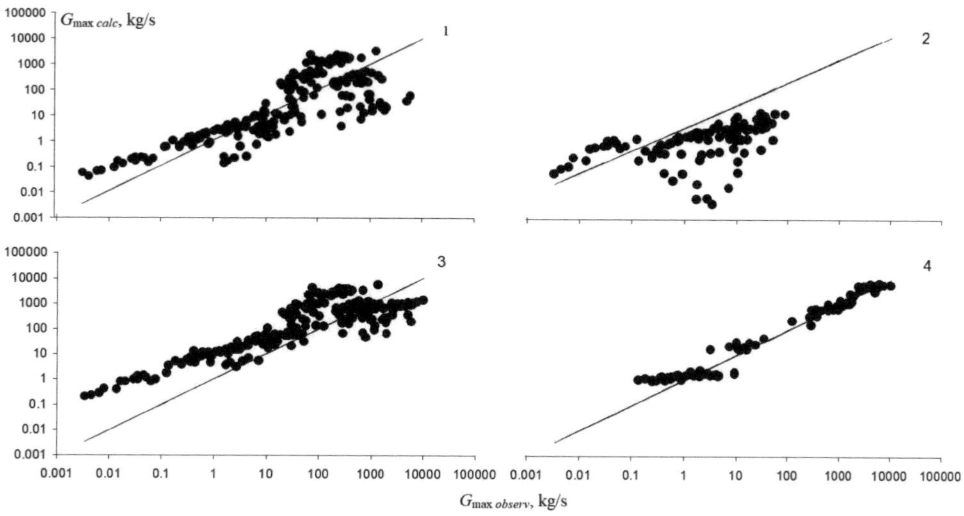

Figure 4.
Observed $G_{max\ observ}$ *and calculated* $G_{max\ calc}$ *flow transport capacity values using Zamarin's formula (Eq. (35))—1, Goncharov's formula (Eq. (36))—2, Bagnold's formula (Eq. (39))—3, and Shmakova's formula (Eq. (25))—4.*

Table 3 shows the calculation data and the results of the calculations for the station Fork Toutle River near Kid Valley. The v and v_{max} values for this station, the only one among all the study stations, showed sufficient proximity. The discrepancy between these values is between 1 and 15%. G_{max} was calculated using three of the formulas (except Goncharov's formula (Eq. (36)) due to non-compliance with the conditions of applicability of this formula for the study river).

As can be seen from **Table 3**, the Shmakova's formula (Eq. (25)) shows the best agreement between the observed and calculated values of the flow transport capacity. The average relative deviation for this formula was 29%. For Zamarin's (Eq. (35)) and Bagnold's (Eq. (39)) formulas, the calculated G_{max} values were significantly lower than the observed values.

Many of the formulas for flow transport capacity and minimum non-silting velocity have been derived experimentally for canals at maximum sediment load. In the case of natural watercourses, however, selecting the formula for the transport capacity and optimizing its parameters is extremely difficult due to the lack of reliable observational data on the maximum possible of sediment load in the flow. For this reason,

Q, m³/s	v_{max}, m/s	v, m/s	B, m	h, m	$G_{max\ observ}$, kg/s	G_{max}, kg/s		
						Zamarin's (Eq. (35))	Bagnold's (Eq. (39))	Shmakova's (Eq. (25))
185	2.9	2.8	56	1.1	5090	110	352	3367
110	2.6	2.4	56	0.9	1883	53	181	1953
123	2.6	2.6	56	0.85	2045	64	269	2417
147	2.6	2.4	59	0.85	1714	69	260	2607
153	2.8	2.4	59	0.98	1978	71	225	2348

Table 3.
Calculation data and calculation results for transport capacity, Fork Toutle River near Kid Valley.

the calculations in this chapter are somewhat tentative. However, the results obtained demonstrate the incompleteness of theoretical research in the study of the transport capacity of rivers. Concurrently, an important advantage of the formulas (Eqs. (17), (25), (28), and (31)) is analytical conclusion from the equation of basic two-phase mass carryover in river flow (Eq. (10)). In this equation, the forces are written not in relation to the water flow, but in relation to a moving solid (the shearing projection of the gravity of the water flow, the retaining projection of the gravity of moving particles, the inertia forces of the water flow and moving particles, the force of the soil resistance to shear). Also in equation (Eq. (10)), the interaction of the water flow, and the bottom is represented by the resistance of the bottom sediment to the tangential load from the flow side. Therefore, formulas (Eqs. (17), (25), (28), and (31)) are based on interrelated calculation of water flow and solid matter and supported by qualitative, but not quantitative characteristic of bottom sediment size. Friction parameters are derived functionally from bottom sediment size categories, which are represented by wide ranges of bottom sediment sizes.

6. Conclusions

Two-phase river flow represents a complicated system of liquid and solid phases and underlying surface interaction. Its main particulars are as follows:

1. Sediment transport in the river flow is, from one hand, predefined by flow hydrodynamics and, from the other hand, impacts to hydraulic variables of flow state;

2. Bottom sediment size is highly variable along the channel and in water flow and depends on water content phase;

3. Resistance on solid boundary of a flow in cross section is not constant within a year;

4. Practice of calculating solid runoff shows insufficient study of the interaction of river flow and river bed.

Existing methods of sediment transport calculations are not always universal and do not fit for flows of any scale and different hydraulic conditions. In this case, sediment transport process in any kind (suspended or bed load) and any degree of flow saturation with solid phase (from clarified to transporting capacity) is the same for all types of channels and for any water content periods. This process is based on the power of the river flow, which determines the amount of solid matter transported. Accordingly, assessment algorithms of any kind sediment discharge (suspended and bed load) shall be the result of theoretic equations describing two-phase river flow hydrodynamics. This means that these algorithms (formulas) shall be fully interconnected each other and follow one from another. And the structure of sediment transport formula shall be fully coordinated with measuring base opportunities. In particular, the average size of bottom sediment, highly variable in cross section, or its quantile values with specified occurrence decrease calculation accuracy. In this case, integral river bed characteristics, such as bottom sediment size categories, are more convenient for operation. Algorithms, provided in this work, can avoid the above deficiencies in a whole and increase calculation accuracy of sediment discharge for different types of rivers.

Acknowledgements

The research was funded by the study № 0154-2019-0003 of the state research plan for the Institute of Limnology RAS.

Author details

Marina Shmakova
Institute of Limnology RAS, Saint-Petersburg, Russia

*Address all correspondence to: m-shmakova@yandex.ru

References

[1] Shmakova MV. Raschety tverdogo stoka rek i zaileniya vodohranilishch [Calculations of Solid River Flow and Reservoir Siltation]. SPb: VVM Publ; 2018. p. 149 (in Russian)

[2] Vanoni VA. Sedimend transportation mechanics: Suspension of sediment. Journal of the Hydraulics Division - Civil Engineering Database. 1963;**89**:45-46

[3] Vanoni VA. Fifty years of sedimentation. Journal of Hydraulic Engineering. 1980;**110**(8):1021-1057

[4] Karasev IF. Ruslovye processy pri perebroske stoka [Channel Processes in Runoff Transfers]. Leningrad: Gidrometeoizdat Publ; 1975. p. 288 (in Russian)

[5] Goncharov VN. Dimnamika rechnyh potokov [River Flow Dynamic]. Leningrad: Gidrometeoizdat Publ; 1962. p. 366 (in Russian)

[6] Zamarin EA. Capacitie Transportiruyushchaya sposobnost' i dopuskaemye skorosti techeniya v kanalah [Transport Capacity and Permissible Flow Velocities in Channels]. Moscow-Leningrad: Gostransizdat Publ; 1951. p. 82 (in Russian)

[7] Soulsby R. Dynamics of Marine Sands. UK: Thomas Telford; 1997

[8] Karaushev AV. Problemy dinamiki estestvennyh potokov [Problems of Natural Flow Dynamics]. Leningrad: Gidrometeoizdat Publ; 1960. p. 393 (in Russian)

[9] Shamov GI. Formuly dlya opredeleniya predel'noj skorosti i raskhoda donnyh nanosov [Formulas for determining speed limit and bottom sediment flow rate]. Trudy GGI [Proceedings SHU]. 1952;**36**(90):3-17 (in Russian)

[10] Taylor GJ. Diffusion by continuous movement. Proceedings of the London Mathematical Society. 1921;**2**(20):196-212

[11] Schmidt W. Der Massenaustausch in frieer Luft und verwandte Erscheinungen. Hamburg: H.Grand Publ; 1925. p. 118

[12] Makkaveev VM. Uchet vetrovogo faktora v dinamike voln i perenosnyh techenij [Consideration of the wind factor in the dynamics of waves and transport currents]. Trudy GGI [Processing SHI]. 1951;**28**(82):3-35 (in Russian)

[13] Rouse H. An analysis of sediment transportation in the light of fluid turbulence. In: Presented Before the Waterways Division at the Annual Meeting of the American Society of Civil Engineers. New York City; January 19 1939

[14] Van Rijn LC. Sediment transport. II: Suspended load transport. Journal of Hydraulic Engineering ASCE. 1984b;**110**(11):1613-1641

[15] Sedaei N, Honarbakhsh A, Mousavi F, Sadatinegad J. Suspended sediment formulae evaluation, using field evidence from Soolegan River. World Applied Sciences Journal. 2012; **19**(4):486-496

[16] Bagnold RA. An approach to the sediment transport problem from general physics. Professional Paper 422-I. U.S.; 1966

[17] Levi II. Dinamika rechnyh potokov [River Flow Dynamic]. Leningrad-Moscow: Gosenergeticheskoe Publ; 1948. p. 224 (in Russian)

[18] Grishanin KV. Dinamika ruslovyh potokov [Channel Flow Dynamics]. Leningrad: Gidrometeoizdat Publ; 1979. p. 428 (in Russian)

[19] Egiazarov IV. Rashod donnyh nanosov [Bed Load Discharge]. Vestnik armyanskoy akademii nauk [Herald of the Armenian Academy of Sciences]. No 4. Erevan; 1950 (in Russian)

[20] Van Rijn LC. Sediment transport. I: Bed load transport. Journal of Hydraulic Engineering ASCE. 1984a;**110**(10): 1431-1456

[21] Meyer-Peter E, Favre H, Einstein A. Neuere Versuchsresultate uber den Geschiebertrieb. Schweiz Bauzeitung. 1934;**103**(13):89-91 (in German)

[22] Meyer-Peter E, Müller R. Formula for bedload transport. In: Proceedings of the 2nd Meeting of the IAHR. Stockholm; 1948. pp. 39-64

[23] Schoklitsch A. Der Geschiebetrieb und die Geschie—befracht. Wasserkraft und Wasserwirtschaft. 1934;**29**(4):37-43 (in German)

[24] Schoklitsch A. Handbuch des wasserbaues. Vol. 1. Vienna: Springem; 1962. pp. 173-177

[25] Gilbert GK. The transportation of debris by running water. Professional Paper 86. Washington, DC: US Geological Survey; 1914. p. 261

[26] Einstein HA. Formula for the transportation of bed-load. Transactions of the ASCE 107. Washington, DC; 1942. pp. 561-597

[27] Einstein HA. The bed-load function for sediment transportation in open channel flows. Technical Bulletin 1026. U.S. Department of Agriculture, Soil Conservation Service; 1950

[28] Velikanov MA. Principle of the gravitational theory of the movement of sediment. Bulletin of the Academy of Sciences of the U.S.S.R. Geophysics Series. 1954;**4**:349-359 (in Russian)

[29] Shen HW, Hung CS. Chapter 14: An engineering approach to total bed material load by regression analysis. In: Shen HW, editor. Proc. Sedimentation Symposium. Highlands Ranch, CO: Water Resources Publications; 1972. pp. 14-1-14-17

[30] Yang SQ, Lim SY. Total load transport formula for flow in alluvial channels. Journal of Hydraulic Engineering. 2003;**129**(1):68-72

[31] Ackers P. Sediment transport in open channels: Ackers and White update. Proceedings of the Institution of Civil Engineers - Water, Maritime and Energy. 1993;**101**:247-249

[32] Ackers P, White WR. Sediment transport: New approach and analysis. Journal of Hydraulics Division, ASCE. 1973;**99**(HY11):2041-2060

[33] Karim MF, Kennedy JF. Menu of coupled velocity and sediment-discharge relations for rivers. Journal of Hydraulic Engineering ASCE. 1990;**116**(8):978-996

[34] Karim MF, Kennedy JF. Computer-Based Predictors for Sediment Discharge and Friction Factor of Alluvial Streams. Iowa Institute of Hydraulic Research, University of Iowa; 1983 [Report No. 242]

[35] Yang CT. Sediment Transport: Theory and Practice. New York: McGraw-Hill; 1996

[36] Yang CT. Unit stream power equation for total load. Journal of Hydrology. 1979;**40**(1):123-138

[37] Engelund F, Hansen E. A monograph on sediment transport in alluvial

streams. Teknisk Forlag. Copenhagen, Denmark. Geological Survey, Reston, VA; 1967

[38] Molinas A, Wu B. Transport of sediment in large sand-bed rivers. Journal of Hydraulic Research. 2001; **39**(2):135-145

[39] Visser Paul J. Application of Sediment Transport Formulae to Sand-Dike Breach Erosion. Communications on Hydraulic and Geotechnical Engineering. Report № 94-3. Faculty of Civil Engineering, Delft University of Technology; 1995

[40] Babkov VF, Bykovskij NI, Gerburt-Gejbovich AV, Tulaev AYA. Gruntovedenie mekhanika gruntov [Soil Science and Soil Mechanics]. Moscow: Dorizdat Publ; 1950. p. 334 (in Russian)

[41] Lamaev BF. Gidrostrujnye nasosy i ustanovki [Hydro-Jet Pumps and Installations]. Leningrad: Mashinostroenie Publ; 1988. p. 256 (in Russian)

[42] Studenichnikov BI. Razmyvayushchaya sposobnost' potoka i metody ruslovyh raschetov [Stream Erosion Capacity and Methods of Channel Calculation]. Moscow: Gosstroyizdat Publ; 1964. p. 183 (in Russian)

[43] Available from: http://pubs.usgs.gov/of/1989/0067/report.pdf

Chapter 3

Assessment of Hydraulic Conductivity of Porous Media Using Empirical Relationships

Abhishish Chandel and Vijay Shankar

Abstract

Flow-through porous media is concerned with the term hydraulic conductivity (K), which imparts a crucial role in the groundwater processes. The present work examines the impact of key parameters i.e., grain size and porosity on the K of four borehole soil samples (Gravelly, Coarse, Medium, and Fine sands) and evaluates the applicability of seven empirical relationships for K estimation. Experimental investigations postulate that an increase in the grain size and porosity value increases the K value. Further, the K values computed using the Kozeny–Carman relationship proved to be the best estimator for Coarse, medium, and fine sands followed by Beyer and Hazen relationships. However, the Beyer relationship had a closer agreement with experimentally obtained value for Gravelly sand. Alyamani and Sen relationship is very sensitive toward the grain-size curve pattern, hence it should be used carefully. Whereas other relationships considered in this study underestimated the K of all samples.

Keywords: empirical relationship, grain size, hydraulic conductivity, porosity

1. Introduction

Water is a vital natural resource, imperative for the existence of all living organisms. Various processes i.e., agricultural processes, groundwater management practices, and environmental quality are influenced by water [1]. Apart from agriculture, some other usages are municipal and industrial water supply, hydroelectric power, forestry, and navigation [2]. Water is available in different forms i.e., surface water, groundwater, ice caps, and glaciers, however, groundwater seems to be a more consistent source of water [3]. Investigation on the computation of hydraulic conductivity of borehole soil samples results in a potential alternative for groundwater monitoring [4].

Initially, Darcy's law defines the term K as the 1-D flow of water through the saturated porous sediments [5]. K is the dominant hydraulic parameter of the porous media used for predicting the movement of fluid through the connecting voids [6]. It has a significant role, in estimating the quantity of seepage through earth dams and levees and conducting stability analysis of earth structures subjected to seepage forces [7]. Saturated *K* of porous media is important for modeling the flow of water in the saturated zone [8, 9]. In the previous studies, it was postulated that the *K* of granular porous media is related to grain size characteristics i.e., d_{10}, d_{20}, d_{50}, and d_{60} [10]. This relationship is very convenient for hydraulic conductivity estimation in the initial stages of aquifer investigation. The representative grain size of porous media from the gradation analysis is helpful in the assessment of *K* values [11]. Various properties influenced the *K* of porous sediments i.e., porosity, structure alignment as well as different properties of fluid such as temperature and viscosity [12].

In groundwater investigations processes, there are many techniques namely laboratory and field methods, and empirical equations are available to estimate the *K* of porous sediments [13]. Precise knowledge of aquifer geometry and boundaries consequences the limited use of field methods [14]. Also, the collection of undisturbed soil samples is a challenging factor concerning the laboratory experimental techniques. Therefore, the computation of *K* values using empirical relationships has been used as a substitute to overcome the issues that occur due to the field and laboratory techniques [15, 16]. Various investigators derive the empirical relationships to compute the *K* value and should be used within particular domains of applicability [17]. The computed *K* values based on different empirical relationships to the similar size of porous sediments can result in different *K* values because the applicability and domains are different for individual empirical relationships [18].

Kasenow [19] analyzed some important empirical relationships on the same porous media and concluded that different *K* values may be obtained. Carrier [20] concluded that the Kozeny–Carman equation is the best estimator of *K* as compared to other empirical relationships. Odong [17] focused on the evaluation of *K* of porous media using empirical relationships and concluded that precise estimation of *K* is based on the Kozeny–Carman equation, however other relationships in the study overestimated the *K* values. Rosas et al. [6] estimated and compared *K* with empirical relationships for 400 samples of sediments with different grain size distributions. Cabalar & Akbulut [21] determined the hydraulic conductivity of sand samples of different shapes and grain sizes and evaluated them with empirical relationships. An M5 model tree was developed to predict *K* based on gradation analysis by Naeej et al. [22]. Ríha et al. [16] evaluated the applicability and reliability of glass beads of different diameters and assessed the *K* of glass beads using empirical relationships. Hong et al. [23] revised the Kozeny–Carman relationship based on effective void ratio and specific surface area and then used it to predict the hydraulic conductivity of the soil.

The literature revealed that different investigators use the existing empirical relationships to estimate the hydraulic conductivity values and vaguely define their applicability boundaries via normal description of materials used without suitable assessment and grain size distribution analysis. The present study has been focused to address this research gap. The main objectives of the study are:

1. To study the influence of key parameters namely grain size and porosity on the *K* of the borehole samples.

2. To determine the flow regime by analyzing the variations between friction factor and Reynolds number.

3. To evaluate the applicability of seven established empirical relationships for *K* estimation.

2. Materials and experimental procedure

In the present study, four representative soil samples were collected during borehole drilling operation at the Una district of Himachal Pradesh in India. Samples (1, 2, 3, and 4) were collected at an interval of 3 m. The collected samples were tested for gradation analysis as per standard procedure to determine different grain sizes i.e., d_{10}, d_{20}, d_{50}, and d_{60} [24]. Samples 1, 2, 3, and 4 containing coarse porous media particles, hence subjected to the dry sieve analysis. A pycnometer test has been conducted on each collected soil sample to determine the specific gravity.

The *K* of soil samples was measured using a constant head permeameter having a diameter of 153 mm and a test length of 46.5 cm as shown in **Figure 1**. Initially, the samples were placed inside an oven at 105°C for about 24 hours for maturing and then added to the permeameter in a completely dry state and were compacted in layers with a rubber mallet. The upper part of the permeameter is connected to a water supply tank, which is situated at a height of 2.5 m above the permeameter, and the lower part is connected to an outlet pipe for discharge measurement. Before determining the *K*, the sample was saturated to maintain a steady flow condition. A constant head was preserved in the manometer pipe, and then the discharge value was measured for a fixed time interval. Five to six measurements were made at different

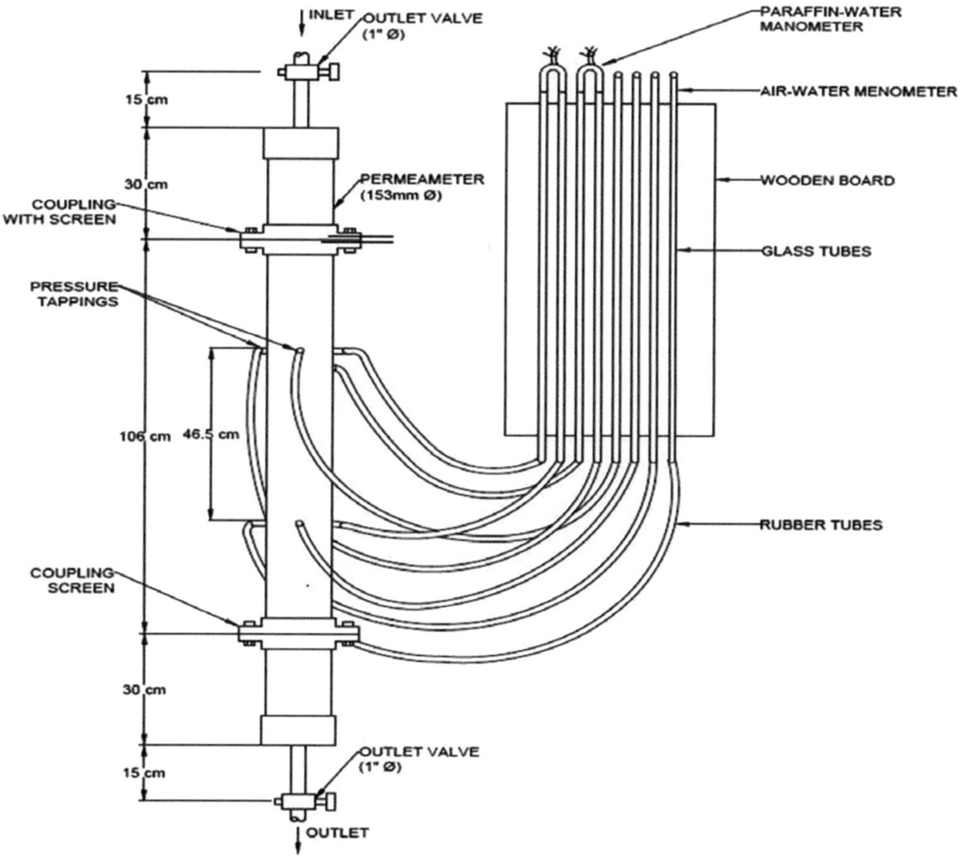

Figure 1.
Hydraulic conductivity measuring setup.

constant heads. The average of the discharge values was taken to determine the K of the sample [25]. The water temperature was measured at the start and the end of the permeameter test. The value of K was calculated by multiplying the flow rate (cm^3/s) by the specimen thickness (cm) and then diving it by the permeameter area (cm^2) times the constant head (cm).

3. Established empirical relationships

From gradation analysis, effective grain diameter and particle uniformity are used to estimate the K using empirical relationships which relate the K value with the size property of porous media. Based on the previous investigations in the field of K computation, Vukovic and Soro [18] formulated a generalized K equation as:

$$K = \frac{g}{\vartheta} * \alpha * f(x) * d_x^2 \tag{1}$$

where, g = gravitational constant, ϑ = Kinematic viscosity, α = sorting coefficient, x = porosity and, d_x = effective grain size. The values of $\alpha, f(x)$, and d_x depend on different procedures used for gradation analysis. Vukovic and Soro [18] postulated a standard equation between uniformity coefficient (U) and porosity as:

$$x = 0.255\{1 + 0.83^U\} \tag{2}$$

$$U = \frac{d_{60}}{d_{10}} \tag{3}$$

where, d_{60} and d_{10} are the grain diameter in (mm).

Several investigators developed different empirical relationships based on the standard equation as mentioned in Eq. (1). **Table 1** represents the seven empirical

Researcher	Relationship	α	Use
Hazen [26]	$K_{Hazen} = \frac{g}{v}.\alpha.[1 + 10\,(x - 0.26)].d_{10}^2$	6×10^{-4}	$0.1\text{ mm} < d_{10} < 3\text{ mm}$ $U < 5$
Slichter [27]	$K_{Slichter} = \frac{g}{v}.\alpha.x^{3.287}d_{10}^2$	1×10^{-2}	$0.01\text{ mm} < d_{10} < 5\text{ mm}$
Terzaghi [28]	$K_{Terzaghi} = \frac{g}{v}.\alpha.\left(\frac{x-0.13}{\sqrt[3]{1-x}}\right)^2 d_{10}^2$	8.4×10^{-3}	Large grain sand
Kozeny [29]-Carman [30, 31]	$K_{K\text{-}C} = \frac{g}{v}.\alpha.\left[\frac{x^3}{(1-x)^2}\right].d_{10}^2$	8.3×10^{-3}	$d_{10} < 3.0$ mm suitable for gravel, sand, and silty soil
Beyer [32]	$K_{Beyer} = \frac{g}{v}.\alpha.\log\frac{500}{U}d_{10}^2$	6×10^{-4}	$0.06\text{ mm} < d_{10} < 0.6\text{ mm}$ $1 < U < 20$
USBR [33]	$K_{USBR} = \frac{g}{v}.\alpha.d_{20}^{2.3}$	4.8×10^{-4}	$U < 5$ Medium grained sand
Alyamani and Sen [34]	$K_{A/S} = \alpha[I + 0.025(d_{50} - d_{10})]^2$	1300	Well distributed sample

U = *uniformity coefficient,* d_{10} *and* d_{50} = *grain size (mm) and,* I = *line intercept in mm formed by* d_{50} *and* d_{10} *with the grain size axis.*

Table 1.
Empirical relationships for hydraulic conductivity estimation.

relationships, which are used in the present work for the computation of *K* values of porous sediments.

4. Results and discussion

Gradation analysis has been conducted on the collected soil samples to compute the gradation characteristics namely grain size, uniformity coefficient, and porosity. The influence of porosity and various grain sizes on the *K* of soil samples was investigated. Further, the experimentally measured *K* values were compared with the values determined via the empirical relationships.

5. Gradation analysis

Initially, the gradation analysis has been performed on the collected soil samples using a mechanical sieve device. **Figure 2** shows the gradation curve for different borehole soil samples. The gradation curve analysis helps to categorize the soil samples based on particle size as shown in **Table 2** [35].

From the gradation curve, grain size at 10%, 20%, 50%, and 60% cumulative weight was determined. The uniformity coefficient, intercept, and porosity values for samples 1, 2, 3, and 4 are given in **Table 3**.

5.1 Variation of hydraulic conductivity with grain size and porosity

From the gradation analysis, the grain size i.e., d_{10}, d_{20}, d_{50}, and d_{60} for all samples were determined. A linear variation between hydraulic conductivity and effective grain size (d_{10}) was observed. From this, it is concluded that as the value of effective grain size increases the hydraulic conductivity also increases. The variation falls on the

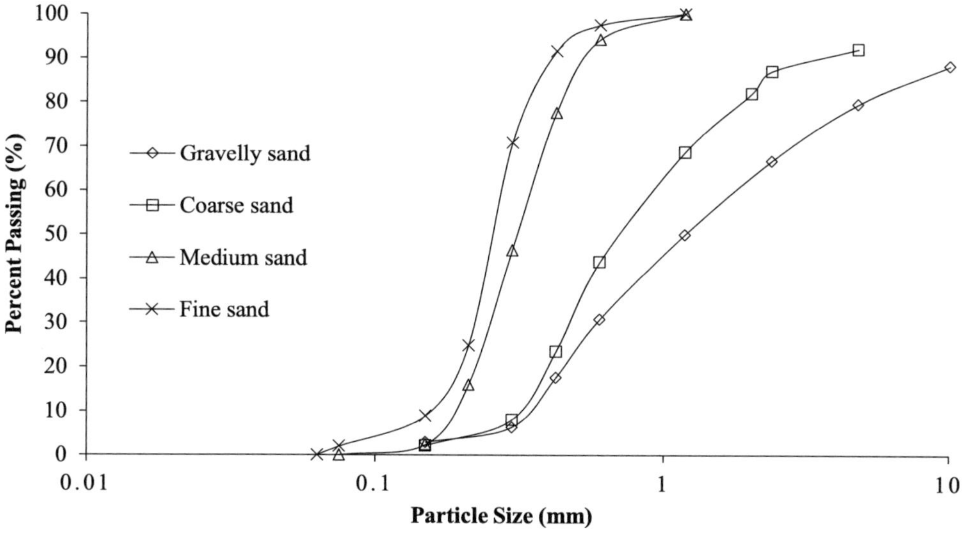

Figure 2.
Gradation curve of collected samples.

Sample	Composition	Classification
1	6% medium gravel, 20% fine gravel, 35% coarse sand, 25% medium sand, and 14% fine sand	Gravelly sand
2	12% fine gravel, 40% coarse sand, 43% medium sand, and 5% fine sand	Coarse sand
3	2% coarse sand, 84% medium sand, and 14% fine sand	Medium sand
4	48% medium sand and 52% fine sand	Fine sand

Table 2.
Soil samples classification based on gradation curve.

Sample – its classification	d_{10} (mm)	d_{20} (mm)	d_{50} (mm)	d_{60} (mm)	I (intercept) (mm)	(U) uniformity coefficient	(x) porosity
1 – Gravelly sand	0.330	0.465	1.160	1.750	0.255	5.30	0.425
2 – Coarse sand	0.316	0.404	0.950	1.380	0.249	4.36	0.412
3 – Medium sand	0.180	0.230	0.390	0.480	0.154	2.67	0.390
4 – Fine sand	0.165	0.190	0.260	0.386	0.135	2.34	0.382

Table 3.
Grain size and other important properties of samples.

same lines for other gain sizes i.e., (d_{20}, d_{50}, and d_{60}) as shown in **Figure 3**. The porosity (x) for all samples was determined and plotted against hydraulic conductivity as shown in **Figure 4**. A straight line obtained between these two parameters indicates that K increases as the porosity value increases, which is in line with the findings of Fallico [36].

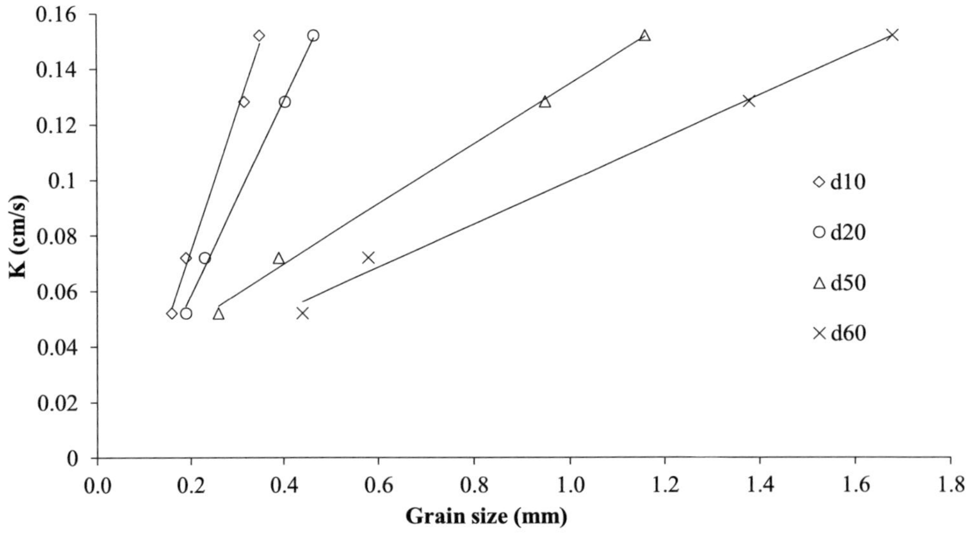

Figure 3.
Variation of hydraulic conductivity with grain size.

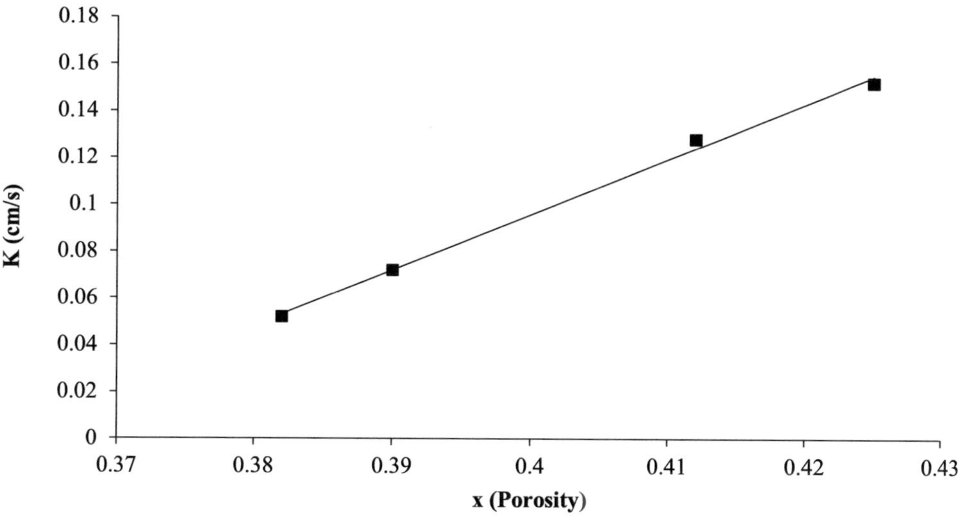

Figure 4.
Variation of hydraulic conductivity with porosity (x).

5.2 Flow regime analysis

To govern the flow regime the variation between dimensionless quantities i.e., friction factor (Fr) and Reynolds number (Re) were studied and plotted on a logarithmic scale. The standard equation to compute the Fr and Re are given as:

$$\mathrm{Fr} = \frac{h_i * g * d_{50} * 2}{U^2} \tag{4}$$

$$\mathrm{Re} = \frac{U d_{50}}{\varphi} \tag{5}$$

where, h_i = hydraulic gradient, d_{50} = average size, g = gravitational constant, U = flow velocity, and φ = fluid kinematic viscosity.

From **Figure** 5, a linear plot between Fr and Re was observed having the Reynolds number value less than 1 [37], which validates the flow regime to be Darcy's or linear regime.

5.3 Computation of *K* using empirical relationships

Initially, the *K* of borehole samples was determined using a constant head permeameter. For Gravelly, Coarse, Medium and Fine sands the obtained *K* value was found to be 0.152, 0.128, 0.072, and 0.052 cm/s respectively. From gradation analysis different parameters i.e., grain size, uniformity coefficient, intercept, and porosity values were determined, which has been used to compute the *K* of all samples using seven empirical relationships. The value of kinematic viscosity i.e., 0.885 mm^2/s derived at a temperature of 27°C was used in the estimation of *K* using empirical relationships. The computed *K* value of all samples using empirical relationships is given in **Table 4**.

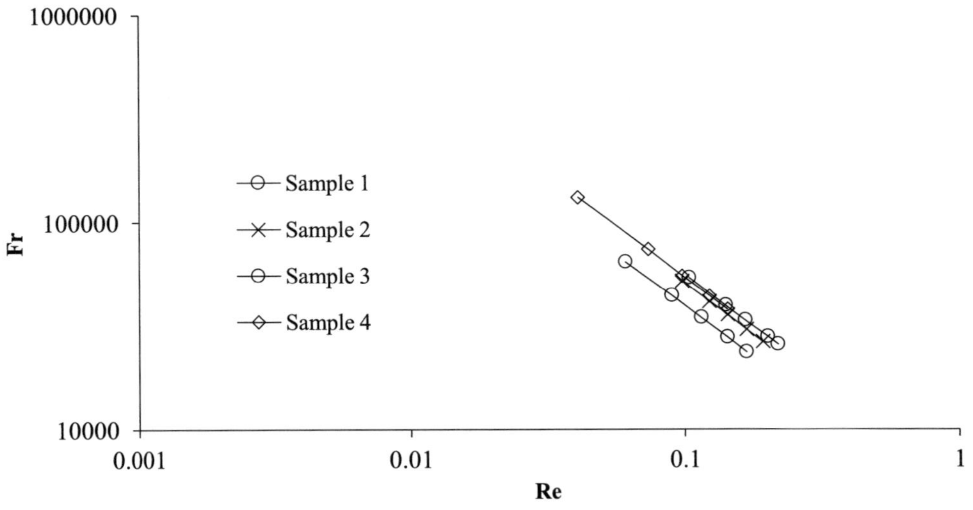

Figure 5.
Plot between Fr and Re of collected soil samples.

Sample – its classification	K_{exp} (cm/s)	K_{Hazen} (cm/s)	$K_{Slichter}$ (cm/s)	$K_{Terzaghi}$ (cm/s)	$K_{K\text{-}C}$ (cm/s)	K_{Beyer} (cm/s)	K_{USBR} (cm/s)	$K_{A/S}$ (cm/s)
1 – Gravelly sand	0.152	NA	0.072	0.093	0.233	0.147	NA	0.114
2 – Coarse sand	0.128	0.167	0.059	0.076	0.132	0.136	NA	0.105
3 – Medium sand	0.072	0.050	0.016	NA	0.064	0.049	0.018	0.038
4 – Fine sand	0.052	0.040	0.013	NA	0.046	0.041	0.012	0.028

NA = not applicable.

Table 4.
Estimation of hydraulic conductivity using empirical relationships.

Hazen and USBR empirical relationships are irrelevant to estimate the *K* value of gravelly sand because the value of uniformity coefficient is greater than 5. For medium and fine sands, the Terzaghi relationship was not used because it is relevant only for large grain sand [14]. Also, the USBR equation applies only to the sizes of medium sand and is thus irrelevant for coarse sand.

Slichter, Terzaghi, and USBR relationships underestimate the *K* values, which is consistent with the findings of Cheng and Chen [38]. Alyamani and Sen relationship results in relatively good prediction for Gravelly and Coarse sands, but it underestimates the *K* for medium and fine sands, because of their poor grading. The *K* values computed using the Kozeny–Carman relationship for coarse, medium, and fine sands have a closer agreement with the measured values followed by Beyer and Hazen relationships as shown in **Figure 6**. The K-C relationship underestimated the *K* value for Gravelly sand because the relationship is not suitable if the grain size distribution has a flat, long tail of fine fraction [20]. Beyer relationship provides better *K* prediction for gravely sand.

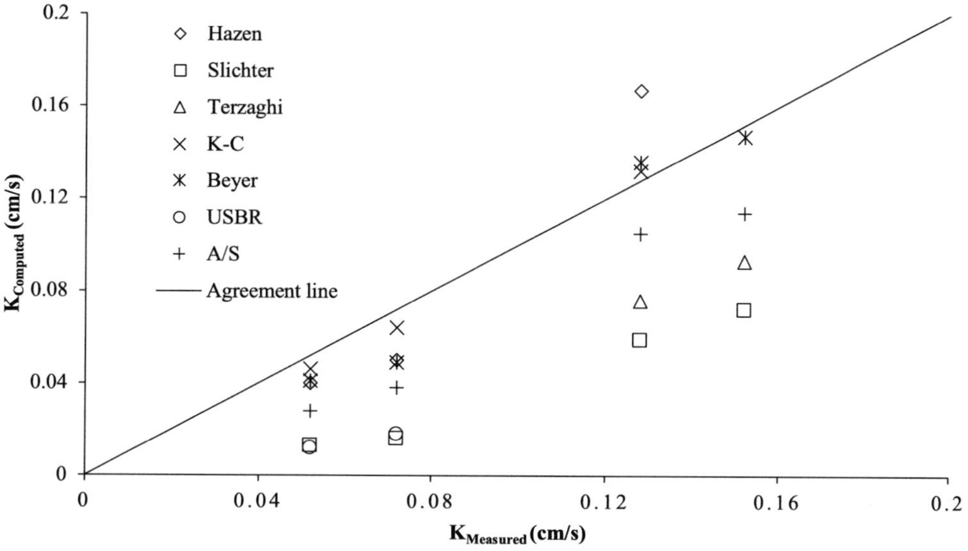

Figure 6.
Comparison of measured and empirically computed hydraulic conductivity.

6. Conclusions

The present work is focused to evaluate seven established empirical relationships for estimating the K of borehole soil samples. Hydraulic conductivity estimation using gradation analysis can also lead to underestimation until the relevant empirical relationship is used. The study examines the impact of key parameters i.e., grain size and porosity on the K value. From the experimental study, it has been observed that with the increase in the grain size and porosity, the K of borehole soil samples increases. The computed K values using the Kozeny–Carman relationship have a closer agreement with the measured values for coarse, medium, and fine sands followed by Beyer and Hazen relationships. Notably, the Beyer relationship provides a better K prediction for Gravelly sand. Alyamani and Sen relationship depends on the grain size curve pattern and should be used carefully. Other relationships i.e., Slichter, Terzaghi, and USBR manifestly underestimate the K of borehole samples.

Author details

Abhishish Chandel* and Vijay Shankar
National Institute of Technology, Civil Engineering Department, Hamirpur, India

*Address all correspondence to: abhishishchandel@gmail.com; vsdogra12@gmail.com

References

[1] Kiran DA, Ramaraju HK. Assessment of water and soil quality along the coastal region of Mangaluru. Journal of Indian Association for Environmental Management (JIAEM). 2019;**39**(1–4): 9-13

[2] Loucks DP, Van Beek E. Water Resource Systems Planning and Management: An Introduction to Methods, Models, and Applications. Springer. Paris: UNESCO; 2017

[3] Alabi AA, Bello R, Ogungbe AS, Oyerinde HO. Determination of groundwater potential in Lagos State University, Ojo; using geoelectric methods (vertical electrical sounding and horizontal profiling). Report and Opinion. 2010;**2**(5):68-75

[4] Anomohanran O. Geophysical investigation of groundwater potential in Ukelegbe, Nigeria. Journal of Applied Sciences. 2013;**13**(1):119-125

[5] Deb SK, Shukla MK. Variability of hydraulic conductivity due to multiple factors. American Journal of Environmental Sciences. 2012;**8**(5):489

[6] Rosas J, Lopez O, Missimer TM, Coulibaly KM, Dehwah AH, Sesler K, et al. Determination of hydraulic conductivity from grain-size distribution for different depositional environments. Groundwater. 2014;**52**(3):399-413

[7] Říha J. Groundwater flow problems and their modelling. In: Assessment and Protection of Water Resources in the Czech Republic. Cham: Springer; 2020. pp. 175-199

[8] Chandel A, Shankar V, Alam MA. Experimental investigations for assessing the influence of fly ash on the flow through porous media in Darcy regime. Water Science and Technology. 2021; **83**(5):1028-1038

[9] Perkins KS, Elango L. Measurement and modeling of unsaturated hydraulic conductivity. In: Hydraulic Conductivity–Issues, Determination and Applications. China: Intech; 2011. pp. 419-434

[10] Sperry JM, Peirce JJ. A model for estimating the hydraulic conductivity of granular material based on grain shape, grain size, and porosity. Groundwater. 1995;**33**(6):892-898

[11] Boadu FK. Hydraulic conductivity of soils from grain-size distribution: New models. Journal of Geotechnical and Geoenvironmental Engineering. 2000; **126**(8):739-746

[12] Omojola AD, Akinpelu SJ, Adesegun AM, Akinyemi OD. A micro study to determine porosity, hydraulic conductivity, permeability and the discharge rate of groundwater in Ondo state riverbeds, southwestern Nigeria. International Journal of Geosciences. 2014;**5**(11):1254

[13] Todd DK, Mays LW. Groundwater Hydrology. New York: John Wiley & Sons; 2005

[14] Ishaku JM, Gadzama EW, Kaigama U. Evaluation of empirical formulae for the determination of hydraulic conductivity based on grain-size analysis. Journal of Geology and Mining Research. 2011;**3**(4):105-113

[15] Cirpka OA. Environmental fluid mechanics I: Flow in natural hydrosystems. Journal of Hydrology. 2003;**283**:53-66

[16] Říha J, Petrula L, Hala M, Alhasan Z. Assessment of empirical formulae for

determining the hydraulic conductivity of glass beads. Journal of Hydrology and Hydromechanics. 2018;**66**(3):337-347

[17] Odong J. Evaluation of empirical formulae for determination of hydraulic conductivity based on grain-size analysis. Journal of American Science. 2007;**3**(3):54-60

[18] Vukovic M, Soro A. Determination of Hydraulic Conductivity of Porous Media from Grain-Size Composition. Littleton, Colorado: Water Resources Publications; 1992 [551.49 V 986]

[19] Kasenow M. Determination of Hydraulic Conductivity from Grain Size Analysis. LLC, Highland Ranch, CO, USA: Water Resources Publication; 2002. p. 83

[20] Carrier WD. Goodbye, hazen; hello, kozeny-carman. Journal of Geotechnical and Geoenvironmental Engineering. 2003;**129**(11):1054-1056

[21] Cabalar AF, Akbulut N. Evaluation of actual and estimated hydraulic conductivity of sands with different gradation and shape. Springerplus. 2016; **5**(1):820

[22] Naeej M, Naeej MR, Salehi J, Rahimi R. Hydraulic conductivity prediction based on grain-size distribution using M5 model tree. Geomechanics and Geoengineering. 2017;**12**(2):107-114

[23] Hong B, Li XA, Wang L, Li L, Xue Q, Meng J. Using the effective void ratio and specific surface area in the Kozeny–Carman equation to predict the hydraulic conductivity of loess. Water. 2020;**12**(1):24

[24] ASTM. Standard D422—Particle-Size Analysis of Soils. PA, USA: West Conshohocken; 2007

[25] ASTM. Standard D2434—Permeability of Granular Soils (Constant Head). PA, USA: West Conshohocken; 2006

[26] Hazen A. Some physical properties of sands and gravels, with special reference to their use in filtration. In: 24th Annual Rep., Massachusetts State Board of Health, Pub. Doc. No. 34. 1892. pp. 539-556

[27] Slichter CS. Theoretical investigation of the motion of ground waters. In: The 19th Ann. Rep. US Geophys Survey. 1899. pp. 304-319

[28] Terzaghi KARL. Principles of soil mechanics. Engineering News-Record. 1925;**95**(19–27):19-32

[29] Kozeny J. Uber kapillare leitung der wasser in Boden. Royal Academy of Science, Vienna, Proceedings, Class I. 1927;**136**:271-306

[30] Carman PC. Flow of Gases Through Porous Media. London: Butterworths Scientific Publications; 1956

[31] Carman PC. Fluid flow through granular beds. Transactions. Institute of Chemical Engineers. 1937;**15**:150-166

[32] Beyer W. On the determination of hydraulic conductivity of gravels and sands from grain-size distributions. Wasserwirtschaft Wassertechnik. 1964; **14**(6):165-169

[33] Mallet C, Pacquant J. Les barrages en terre. Paris: Editions Eyrolles; 1951. p. 345

[34] Alyamani MS, Şen Z. Determination of hydraulic conductivity from complete grain-size distribution curves. Groundwater. 1993;**31**(4):551-555

[35] ASTM. Standard practice for classification of soils for engineering

purposes (unified soil classification system). In: Annual Book of ASTM Standards. West Conshohocken, PA: ASTM, International; 2010

[36] Fallico C. Reconsideration at field scale of the relationship between hydraulic conductivity and porosity: The case of a sandy aquifer in South Italy. The Scientific World Journal. 2014;**2014**: 1-15

[37] Chandel A, Shankar V. Evaluation of empirical relationships to estimate the hydraulic conductivity of borehole soil samples. ISH Journal of Hydraulic Engineering. 2021:1-10

[38] Cheng C, Chen X. Evaluation of methods for determination of hydraulic properties in an aquifer–aquitard system hydrologically connected to a river. Hydrogeology Journal. 2007;**15**(4): 669-678

Chapter 4

Study of Polydisperse Particulate Systems with a 'Direct-Forcing/Fictitious Domain' Method

Romuald Verjus and Sylvain S. Guillou

Abstract

Natural sediments responsible for the morphodynamic of the estuaries and coast are of different sizes and densities. Some are cohesive and some are non-cohesive. The transport in suspension and their sedimentation of such a polydisperse suspension are different than the ones for a monodisperse suspension. A fully resolved model based on the Direct-Forcing/Fictitious Domain method (DF/FD) was developed and applied to simulate settling of monodisperse particles in a water column. The behaviour of the suspension corresponds qualitatively to experimental results and average settling velocities follow a Richardson-Zaki type law. Then the model is applied to the sedimentation of suspension composed of particles of three diameters. The segregation of the bed is obtained naturally. The excess pore pressure is drawn and compared with the theory.

Keywords: polydisperse particulate flows, sedimentation, direct numerical simulation, modelling

1. Introduction

Estuaries are places where silting-up events usually occur. Different processes are the root of particle sedimentation on the bed, which contributes to the filling process of estuaries. Flocculation processes (aggregation and break up), erosion, deposition and compaction are major phenomena that must be considered in order to predict the long term behaviour of estuarine sedimentary dynamic. So, in sedimentary transport, as well as other problems with particulate flows, the complex dynamics is not well described [1].

Most of the study had been realised at macro-scales, due to the huge number of particles that constitute these particulate systems, where the fluid and the particles can be considered as a mixture. So it is of single-phase models at large scale without [2] or with sedimentation consolidation model [3]. They can simulate the behaviour of large-scale systems such as estuaries or ports. At smaller scale, Euler–Euler two-phase flow models have been used with success [4–10]. The predictive ability and accuracy of theses codes depend essentially on the quality of the closure equations that model phenomena at smallest scales. Processes such as drag or lift forces acting on a suspension are not well established, and there is clearly a need to better understand smallest-scale phenomena in order to better formulate these constitutive relations.

Direct numerical simulation models for particulate flows have received a great attention for 20 years. In these methods, the fluid phase is governed by the Navier–Stokes equations, and the rigid inclusions are governed by Newton's laws. The flow field around each particle is resolved, and hydrodynamic interactions are results of simulations. Such a model was proposed by Glowinski et al. [11], Peskin [12], Yu and Shao [13], Wachs [14].

Natural sediments responsible for the morphodynamic of the estuaries and coast are of different sizes and densities. Some are cohesive and some are non-cohesive. The transport in suspension and their sedimentation of such a polydisperse suspension are different than the ones for a monodisperse suspension [15]. Thus, the complex behaviour of such particulate systems is not well understood and not really explored by numerical investigations.

We propose here to examine the sedimentation of a polydisperse suspension by the use of a two-dimensional fully resolved model that we developed and validated with monodisperse suspensions in water column. Section 2 briefly describes the mathematical and numerical background of the model. Section 3 presents a validation of the model. Simulation results on the sedimentation of monodisperse and polydisperse particle systems are discussed in Section 4. Finally, Section 5 presents the conclusions and perspectives of the work.

2. Mathematical and numerical background

In the direct numerical simulation models for particulate flows, the fluid phase is governed by the Navier–Stokes equations, and the rigid inclusions are governed by Newton's laws. The flow field around each particle is resolved, and hydrodynamics interactions are results of simulation. The model presented here is based on the 'Direct-Forcing/Fictitious Domain' method of Yu and Shao [13] and the Direct Numerical code of Guillou and Makhloufi [16]. Herein we briefly recall the basic equations with special closure laws. More details can be found in Verjus [17] and Verjus et al. [18].

2.1 Fictitious domain formulation

Let consider a Newtonian fluid and one rigid cylindrical particle. Let $P(t)$ represent the domain of the Particule and Ω the whole domain including the fluid and the particle. Suppose that the particle density, volume and moment of inertia, translational velocity and angular velocity are ρ_s, V_s, J_s, U and ω, respectively. The fluid density and viscosity are ρ_f and μ. Let us introduce scale parameters: L_c for length, U_c for velocity, L_c/U_c for time, $\rho_f U_c^2$ for the pressure and $\rho_f U_c^2/L_c$ for the pseudo body force. Then, the dimensionless fictitious domain equations can be written as Yu and Shao [13]:

$$\frac{\partial \mathbf{u}}{\partial t} + \mathbf{u}.\nabla\mathbf{u} = \frac{\nabla^2\mathbf{u}}{\mathrm{Re}} - \nabla p + \boldsymbol{\lambda} \text{ in } \Omega, \quad (1)$$

$$\mathbf{u} = \mathbf{U} + \boldsymbol{\omega} \times \mathbf{r} \text{ in } P(t), \quad (2)$$

$$\nabla.\mathbf{u} = 0 \text{ in } \Omega, \quad (3)$$

$$(\rho_r - 1)V_d\left(\frac{d\mathbf{U}}{dt} - \frac{1}{Fr}\frac{\mathbf{g}}{g}\right) = -\int_P \boldsymbol{\lambda} dx, \tag{4}$$

$$(\rho_r - 1)\frac{d(\mathbf{J_d}\cdot\boldsymbol{\omega})}{dt} = -\int_P \mathbf{r}\times\boldsymbol{\lambda} dx, \tag{5}$$

where $\boldsymbol{u}$ represents the fluid velocity, p the fluid pressure, $\boldsymbol{\lambda}$ the pseudo body force, $\boldsymbol{U}$ and $\boldsymbol{\omega}$ are respectively the velocity and angular velocity of the particle, $\boldsymbol{r}$ is the position vector with respect to the particle mass centre, Re is the Reynolds number defined by $\rho_f U_c L_c/\mu$, Fr is the Froude number defined by U_c^2/gL_c, ρ_r is the particle-fluid density ratio defined by ρ_s/ρ_f, J_d is the dimensionless moment of inertia defined by $J_s/\rho_s L_c^4$ and V_d the dimensionless volume defined by V_s/L_c^2.

2.2 Numerical method

A fractional-step method is used in order to decouple the problem described by the Eqs. (1)–(5). For a time instant, three major steps are written: one for the prediction of the fluid velocity (and pressure) (Eqs. (6) and (7)); one for the calculation of the particle's motion and the interaction with the fluid (Eqs. (7)–(9)); and the last one to correct the fluid's velocity (Eq. (10)). The present model is an extension of the SUDRES code [16] for the DNS of the fluid flow in two dimensions. A Finite Difference Method on a staggered grid and a projection technique are used to solve the fluid problem. Finally, the time marching-algorithm used is as follows:

- Step 1: Calculation of the fluid velocity and pressure with a projection method

$$\frac{\mathbf{u}^* - \mathbf{u}^n}{\Delta t} = -\nabla^* p - \frac{1}{2}\left(3(\mathbf{u}.\nabla\mathbf{u})^n - (\mathbf{u}.\nabla\mathbf{u})^{(n-1)}\right) + \frac{1}{2}\left\{\frac{1}{\mathrm{Re}}\nabla^2\mathbf{u}^n + \frac{1}{\mathrm{Re}}\nabla^2\mathbf{u}^*\right\} + \boldsymbol{\lambda}^n \tag{6}$$

$$\nabla\cdot\mathbf{u}^* = 0 \tag{7}$$

- Step 2: Calculation of the particle subproblem for $\mathbf{U}^{n+1}$, ω^{n+1}, $\mathbf{u}^{n+1}$ and the body force $\boldsymbol{\lambda}^{n+1}$, with:
  - Step 2.1: Calculation of $\mathbf{U}^{n+1}$, ω^{n+1} depending on $\mathbf{u}^*$ and $\boldsymbol{\lambda}^n$

$$\rho_r V_d \frac{\mathbf{U}^{n+1}}{\Delta t} = (\rho_r - 1)V_d\left(\frac{\mathbf{U}^n}{\Delta t} - \frac{1}{Fr}\frac{\mathbf{g}}{g}\right) + \int_P \left(\frac{\mathbf{u}^*}{\Delta t} - \boldsymbol{\lambda}^n\right)dx \tag{8}$$

$$\rho_r \frac{\mathbf{J_d}\cdot\boldsymbol{\omega}^{n+1}}{\Delta t} = (\rho_r - 1)\frac{\mathbf{J_d}\cdot\boldsymbol{\omega}^n}{\Delta t} + \int_P \mathbf{r}\times\left(\frac{\mathbf{u}^*}{\Delta t} - \boldsymbol{\lambda}^n\right)dx \tag{9}$$

  - Step 2.2: Calculation of $\boldsymbol{\lambda}^{n+1}$ depending on $\mathbf{U}^{n+1}$, ω^{n+1}, $\mathbf{u}^*$ and $\boldsymbol{\lambda}^n$

$$\boldsymbol{\lambda}^{n+1} = \frac{\mathbf{U}^{n+1} + \boldsymbol{\omega}^{n+1}\times\mathbf{r} - \mathbf{u}^*}{\Delta t} + \boldsymbol{\lambda}^n \tag{10}$$

- Step 3: Correction of the fluid velocity $\mathbf{u}^{n+1}$ regarding $\boldsymbol{\lambda}^{n+1}$ and $\boldsymbol{\lambda}^n$

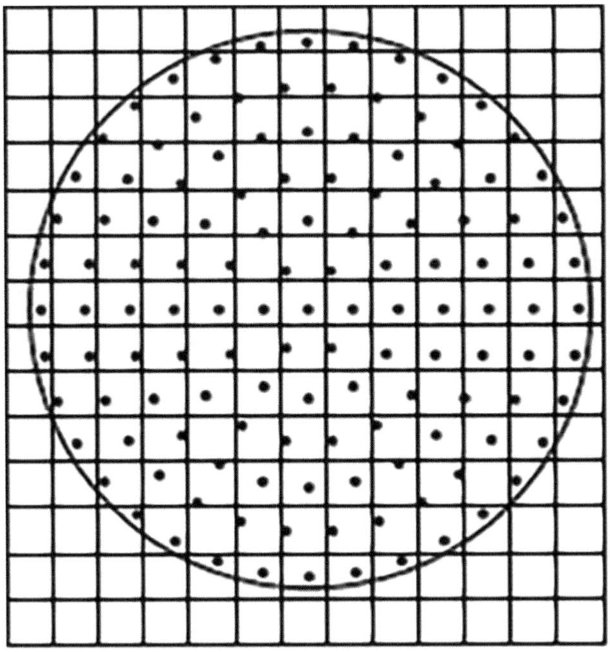

Figure 1.
Meshes of the fluid and solid domains.

$$\mathbf{u}^{n+1} = \mathbf{u}^* + \Delta t\left(\boldsymbol{\lambda}^{n+1} - \boldsymbol{\lambda}^n\right) \tag{11}$$

The discretization of the particle (Lagrangian mesh associated to the particle) is still an open question. In the Distributed Lagrangian Multiplier Method, different techniques are used to discretize the particle [19], but the Collocation Element Method is often used. In the present study, a structured meshing strategy is used to discretize the particle (**Figure 1**). So for the solid problem, the particle is dicretized with the Collocation Point Method. The arrangement of the Lagrangian points is presented in Yu and Shao [13]. Special attention is dedicated to the space between particle nodes: it is greater than the fluid nodes space as suggested by Glowinski et al. [11]. Bilinear interpolations are used to interpolate quantities defined on the fluid's mesh (Eulerian) on the particle's mesh (Lagrangian) and vice versa. A trapezoidal rule is used to perform integrals. The collision strategy employed in this code is based on a normal repulsive force acting when two particles are too close each other [11].

3. Validation

The numerical model is validated by three cases: the simulation of the flow around a fixed particle, the sedimentation of one particle and the sedimentation of two particles.

3.1 Flow around a fixed particle

The first test case is the fixed cylinder (circular particle) in Poiseuille flow. A cylinder is placed at the centre of a channel of width $L = 4D$, where D is the diameter

DOI: http://dx.doi.org/10.5772/intechopen.104654

of the cylinder. The length of the channel is $L = 25D$. Several meshes were tested for $h = D/16, D/32, D/64$. The time step varies from 0.0005 to 0.005 depending on the Reynolds number and the CFL condition. The particular Reynolds number varies from 0.5 to 100. The stream lines and the vorticity contours for different Re_p are presented in **Figure 2**. For $Re_p = 0.5$, the flow is symmetric. For higher Re_p, the inertia becomes important until the apparition of eddies, and at last the periodic detachment of it. **Table 1** presents the drag coefficients for different Re_p and meshes. The values are very close to the ones found by Yu and Shao [13]. The results are better for finer meshes. So the flow dynamic around the cylinder and the interaction with the cylinder are well reproduced.

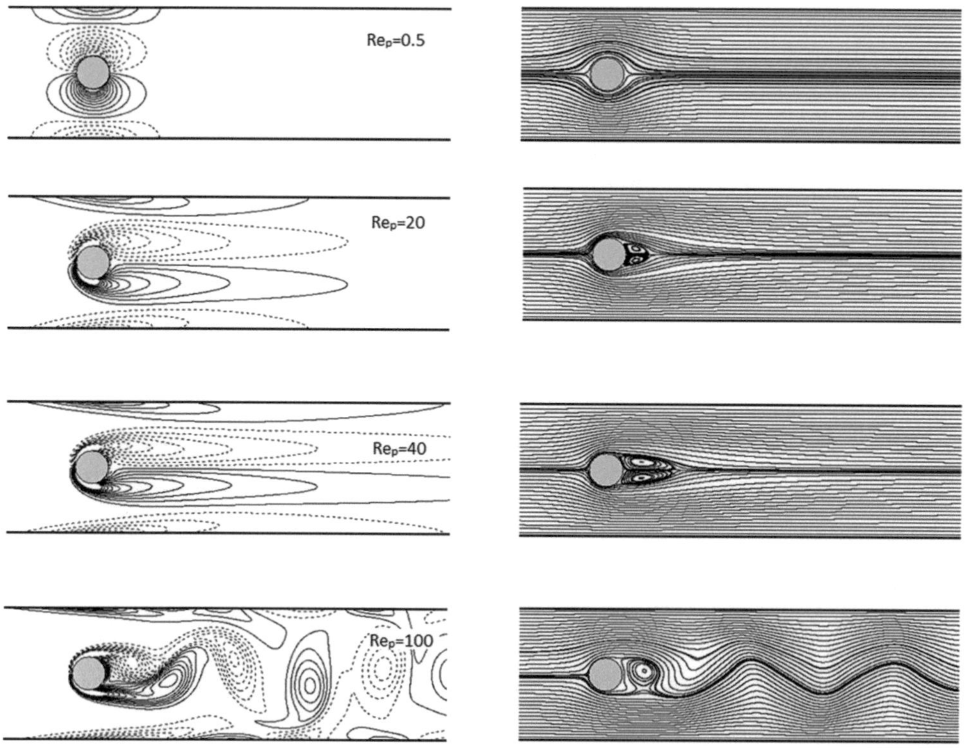

Figure 2.
Flow past a circular cylinder (vorticity on the left and stream line on the right) in a channel (width/diameter = 4) for Re_p = *(0.5, 20, 40, 100).*

Rep	0.5	1	5	10	20	0	100
Present $h = D/16$	93.00	46.68	10.28	6.05	3.96	2.84	2.16
Present $h = D/32$	90.34	46.68	10.01	5.89	3.85	2.75	2.08
Present $h = D/64$	89.11	44.73	9.88	5.83	3.80	2.72	2.06
Yu and Shao [13]	87.95	44.16	9.95	5.75	3.76	2.68	

Table 1.
*Drag coefficients for several Reynolds numbers and for several mesh sizes (*h = D/16, D/32, D/64*).*

3.2 Sedimentation of one particle

The second test case concerns the sedimentation of circular particle in a fluid at rest. The gravity is along the channel axis. The domain and numerical parameters are identical to the ones used for the first case. The Froude and Reynolds numbers are based on an estimation of the settling velocity U_c given by Happel and Brenner [20] at very low Reynolds number (Eq. (12)). It depends on the cylinder diameter, the viscosity, the gravity constant and the density of the fluid and particles and a constant K given by (Eq. (13)) with $L^* = L/D$. Simulations for $Re_p = 0.1$, $L^* = 4$ and $\rho_p/\rho_f = 1.1$ were performed and presented in **Figure 3**. The particle accelerates first and then reaches a constant velocity very close to the theoretical value of U_c ($Re_p = 0.1$) given by Eq. (12).

$$U_C = \frac{D^2}{16K\mu}\left(\rho_p - \rho_f\right)g \tag{12}$$

$$K = \frac{1}{\ln\left(L^*\right) - 0.9157 + 1.7244\left(L^*\right)^{-2} - 1.7302\left(L^*\right)^{-4} + 2.4056\left(L^*\right)^{-6} - 4.5913\left(L^*\right)^{-8}} \tag{13}$$

3.3 Sedimentation of two particles

This test concerns the sedimentation of two particles in a close domain. The simulations for $Re_p = 1$, $L = 8D$, and $\rho_p/\rho_f = 1.1$ were performed and presented on **Figure 4**. The cylinders are located at $2D$ from the axis and their centres are distant of $2D$ on the z-axis. The domain is of size $8D$*$160D$ covers with a uniform mesh of size $h = D/16$.

This configuration corresponds to the experiment of Jayaweera and Mason [21]. As described by Feng et al. [22], the trailing cylinder accelerates in the wake of the leading one and then turns around this one until the centre of the cylinders is horizontally aligned. Afterwards, the two particles move apart on the horizontal direction.

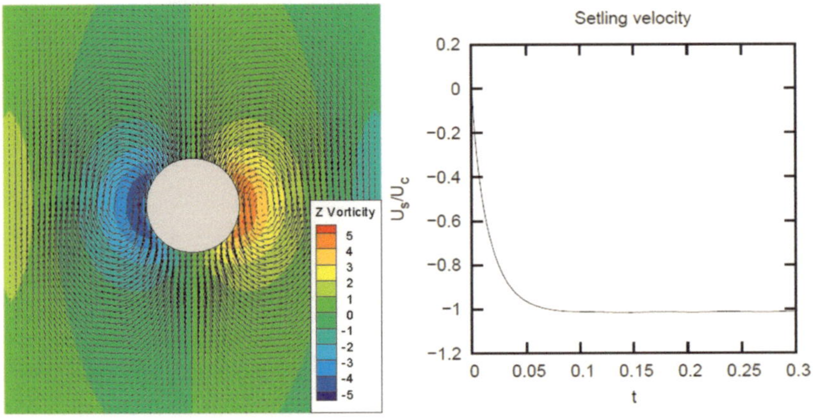

Figure 3.
Sedimentation of a circular particle ($Re_p = 0.1$)*: (a) Isocontour of vorticity at* t = 0.25*; (b) time development of the settling velocity in a vertical channel.*

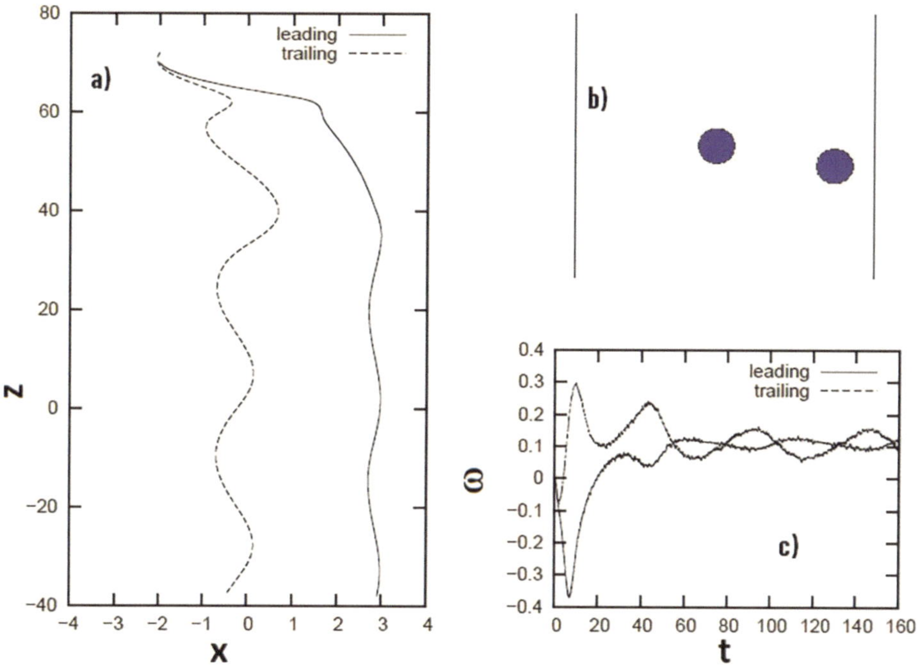

Figure 4.
Sedimentation of two particles: (a) trajectories of the particles; (b) equilibrium positions of the particles; (c) time evolution of the dimensionless rotation velocity.

This phenomenon is found in the simulation. As Feng et al. [22] showed the trailing particle oscillates around the channel axis, whereas the leading one oscillates with smaller amplitude around an axis close to wall. **Figure 4c** shows that the rotation of the particles is accounted for. The rotational velocity is quite small but not nil. More details about the chaotic sedimentation of two particles at low Reynolds number were presented in Verjus et al. [17].

4. Numerical study

4.1 Sedimentation of a mono-disperse suspension

In this section, the sedimentation of a great number of particles is considered. The settling velocity depends on the number of particles. The more there are, smaller is the velocity. In such a case, the terminal velocity formulation is close to the one proposed by Richardson-Zaki [23]: $U_c = U_{ct}(1 - \varphi)^5$, where U_{ct} is the terminal setting velocity of one particle and is the solid volume fraction.

The problem concerns the sedimentation of N_t particles in a close domain. The particles are the same as the previous test ($\rho_p/\rho_f = 1.1$, $Re_p = 0.1$). The width is as $L = 22D$, and the height on which is the initial homogeneous suspension is $H_0 = 88D$ (for the first simulations, then it will be $176D$). The mesh size is $h = D/16$. The time evolution of suspension is presented in **Figure 5**. After some instants the initial structure is destabilised, three domains are visible: the first one conserves the initial concentration, the second one which is close to the bottom has a solid volume fraction

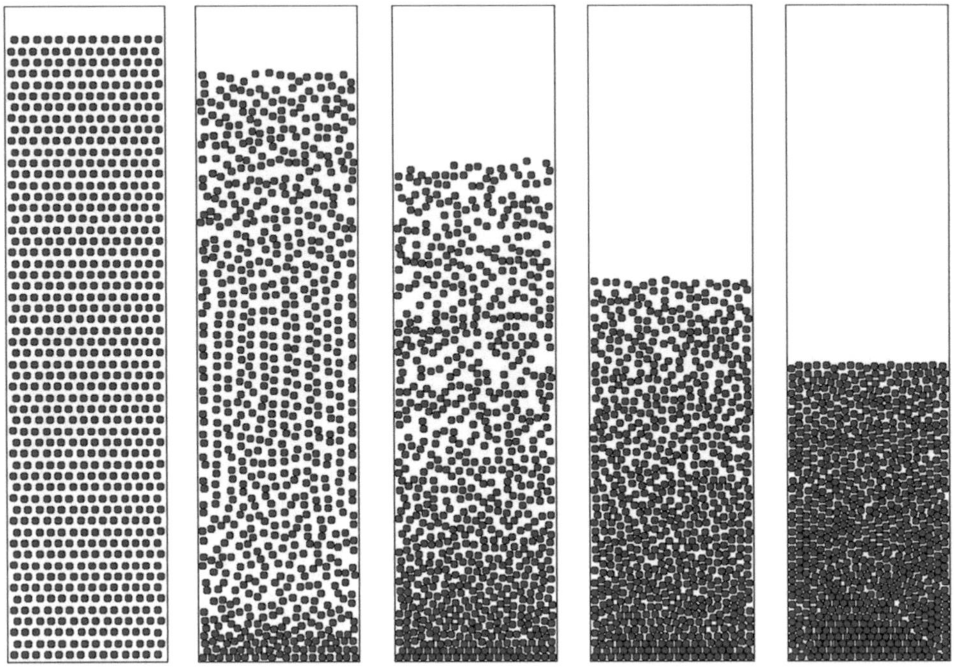

Figure 5.
Snapshot of the sedimentation of an initially homogeneous suspension of 785 particles (L = 22D, Ho = 88D, the initial solid volume fraction is ϕ = 0.32).

close to the saturation one (about 0.7), and then in the third part intermediate solid volume fraction is present (clearly visible here).

A comparison with macroscopic parameter such as the settling velocity of Richardson-Zaki can be made. But a control surface (volume in 3*D*) on which meaning operation will be made must be defined. So the width of this surface is the length *L* (width of the domain) and the height is *H*. Each control volume contains *N* particles. Spatial meaning and temporal operators can be defined as (*T* is the period on which the control volume has a solid volume fraction close to the initial value):

$$\tilde{U}(t) = \frac{1}{N}\sum_{1}^{N} U_i(t) \qquad \overline{U}(t) = \frac{1}{T}\int_{0}^{T} \tilde{U}(t)\mathrm{d}t \tag{14}$$

In **Figure 6** (left), the clear water interface is drawn for several solid volume fractions. At the beginning of the motion, the interface settles with a constant velocity and then reaches a regime for which the evolution slows down. This is hindered settling regime. After a while, there is no motion. It corresponds to the gel point. The hindering regime depends on the number of particles, so its time duration is longer for the higher initial solid volume fraction. The right picture presents the evolution of the dimensionless settling velocity versus the solid volume fraction (0.026, 0.058, 0.103, 0.136, 0.233 and 0.32 which correspond to 136, 306, 544, 850, 1204 and 1666 particles) for *H*0 = 176*D*. The control volume is placed at the centre of the channel. Qualitatively, the settling velocities resulting from the simulations follow the curve provided by the Richardson-Zaki formula with an exponent of 5. But the best fit of the

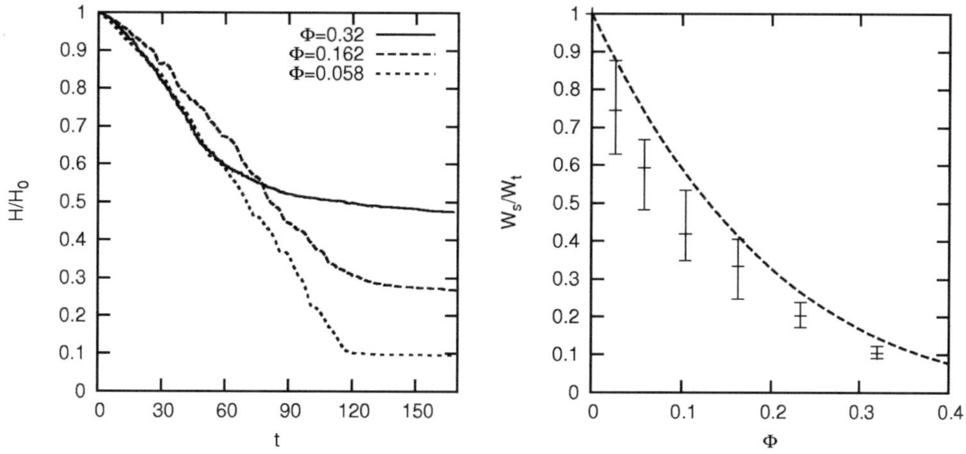

Figure 6.
Left: Time evolution of the clear water-suspension interface (settling curve) for three initial solid volume fractions 0.32, 0.162, 0.058 (L = 22D). Right: evolution of the ratio of the calculated settling velocity (W$_s$) by the one of one particle (W$_t$) versus the solid volume fraction. The dash line corresponds to the settling velocity calculated with the Richardson-Zaki formula and an exponent of 5.

numerical results is obtained for an exponent of 7.2 ± 0.6. That difference can be due to the 2D-simulation, whereas the Richardson-Zaki formula is made for spherical particles.

4.2 Sedimentation of a poly-disperse suspension

In this section, we consider the sedimentation of a poly-disperse suspension. The simulations are two dimensional, and particles are represented by discs with three different diameters: $d_1 = d_{50}\ 2/3$, $d_2 = d_{50}$ and $d_3 = d_{50}\ 4/3$. The mean diameter is noted d_{50}. All the particles have the same ratio of density $\rho_p/\rho_f = 1.5$. The computational domain is closed and of sizes $28\ d_{50} \times 84 d_{50}$ respectively in the horizontal and vertical directions. Initially, the fluid and the particles are at rest with a homogeneous distribution as in **Figure 7**. The particulate Reynolds number $Re_p = W_s.d_{50}/\nu$ is equal to *0.1*. Re_p is based on the settling velocity obtained with the Richardson-Zaki [23] formula. The initial solid volume fraction is 0.107 which corresponds to *300* particles. The time step is fixed at *0.001*, and the fluid mesh is 673 × 2017 nodes which corresponds to a ratio of $h = d_{50}/16$, where h is the fluid mesh size.

Figure 7 presents the motions of particles and the fluid flow for various instants. At the beginning of simulation, the suspension is homogeneous in the domain. After few seconds, most of smallest particles are expulsed upwards, whereas the two other kinds of particles fall on the bottom. However, the biggest particles fall faster than medium ones and then reach the bottom first. In the suspension, two different areas appear: one which is essentially composed of smallest particles in the upper part and a second one constituted with a non-homogeneous mixture of small and medium particles. The bed is constituted in majority of big particles. At the end of simulation, particles are fully segregated, and the bed is composed of three layers filled mainly with respectively one class of particle.

Figure 8 presents the vertical evolution of the net pressure at different instants. This pressure corresponds to the total pressure less the hydrostatic pressure in that

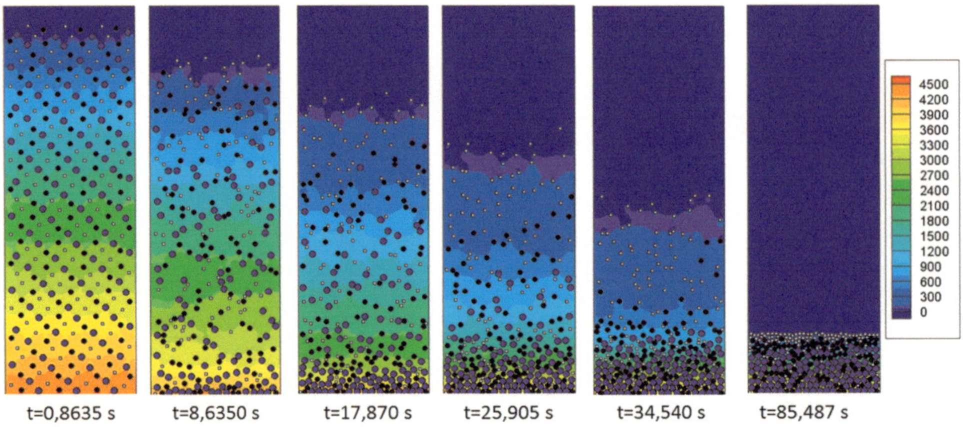

Figure 7.
Snapshots of the sedimentation of tri-disperse particles in a closed box (grey particles for diameter d_1*, black ones for diameter* d_2 *and blue ones for diameter* d_3*). In this simulation the parameters are:* $Re_p = 0.1$*;* $width/d_{50} = 28$ *and the initial volume fraction of 0.107. The contour presents the net pressure (unity:* 10^{-2} *Pa).*

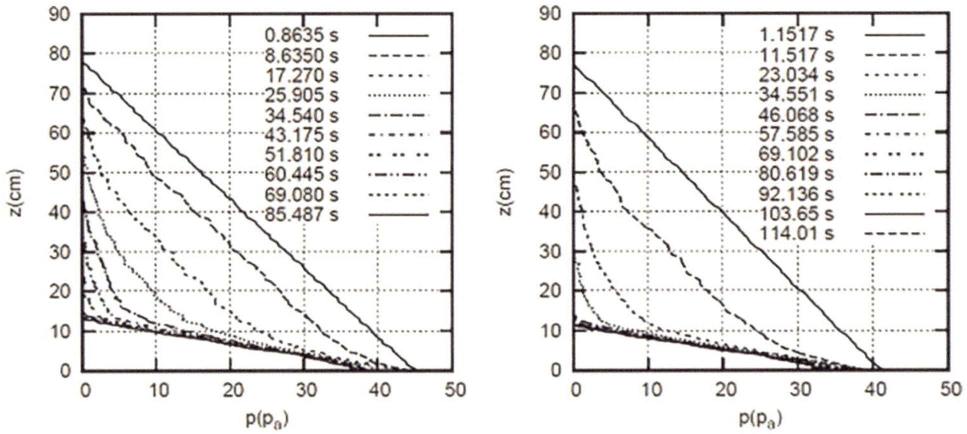

Figure 8.
Net pressure versus height at different instants. Right figure is the simulation with the parameters presented in the article. Left figure, the simulation case differs by the number particles by species (114 particles of diameter d_1*, 112 particles of diameter* d_2 *and 74 particles of diameter* d_3*) which leads to* $d_{50} = 1.33$*.*

case. At the beginning of the simulation, the pressure gradient is constant. Two areas appear at instant $t = 25.905$ s. Surprisingly, these two areas seem to be linear, and we note that the highest-pressure gradient is in the region near the lower wall. At the end of the simulation, the pressure gradient is constant. In mixture theory (e.g. [24]), the net pressure gradient is calculated with:

$$-\frac{dp}{dz} \approx \left(\rho_p - \rho_f\right) \Phi g \tag{15}$$

Where Φ is the solid volume fraction. The pressure gradient obtained with the code for the first time instant is −0.56 Pa/cm and that calculated with (Eq. (15)) is −0.52 Pa/cm. At the end of simulation, when the particles are approximately fixed, the pressure gradient tends to −3.21 Pa/cm. With Eq. (15) and the maximal volume

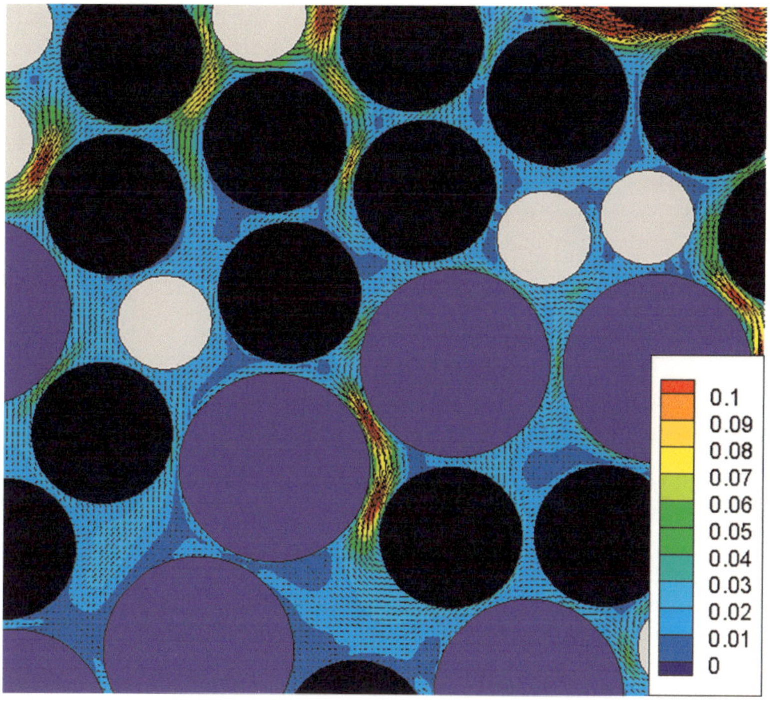

Figure 9.
Particles arrangement in the bed at the end of the simulation. Velocity vectors and contours are plotted (unity: cm/s).

fraction for discs (0.7), the pressure gradient is equal to −3.43 Pa/cm. Results of simulations, with other initial volume fraction confirm this good agreement and seem independent of the polydispersivity. The small difference between Eq. (15) and the results obtained with the code is due to the fact that the volume fraction we calculated depends on the total volume we choose to calculate it. For example, at the beginning of the simulation, the volume fraction is calculated with the total volume of the domain. If only the volume where the particles are located (which give $\Phi = 0.112$) is considered, the accordance with Eq. (15) is better (pressure gradient is equal to −0.55 Pa/cm). Simulations with more particles would reduce the error we made on the calculation of the volume fraction. Moreover, at the end of simulation we have taken the maximal volume fraction. In our cases, the total blockage never appends. Particles in close contact never touch, due to the collision strategy. The flow in the pore remains possible (Cf. **Figure 9**).

5. Conclusions

A fully resolved model based on the Direct-Forcing/Fictitious Domain method (DF/FD) was developed to study the sedimentation of particles. It was validated on simple cases and applied to the sedimentation of mono-disperse particles. The macroscopic behaviour of the solution corresponds to experimental results and average settling velocities follow a Richardson-Zaki type law. The application of the model to a poly-disperse suspension leads to the apparition of a natural segregation of the particles on the bed. As the pressure is computed by the model, the pressure between the grains can be extract. So the excess pore pressure along the vertical is extract. It appears that at the end of the simulation, but not of the rearrangement of the bed, the pressure profile approaches the law proposed by Guazzelli and Morris [24].

The results presented here are limited to low Reynolds numbers, so we will explore other range of Reynolds numbers. The simulation of sedimentation of particles with complex shapes is possible, but a supplementary work is necessary concerning the collision strategy.

Acknowledgements

We would like to thank the Syndicat Mixte du Cotentin for financing the computational resources.

Author details

Romuald Verjus and Sylvain S. Guillou*
Normandy University, University of Caen Noamndie, LUSAC,
Cherbourg-En-Cotentin, France

*Address all correspondence to: sylvain.guillou@unicaen.fr

References

[1] Guillou S, Thiébot J, Chauchat J, Verjus R, Besq A, Nguyen DH, Pouv KS. The filling dynamics of an estuary: From the process to the modelling."Sediment Transport in Aquatic Environments". Ed. A. Manning. London, UK: IntechOpen; 2011. Chap. 6, p. 125-146. ISBN 978-953-307-586-0. http://www.intechopen.com/articles/show/title/the-filling-dynamics-of-an-estuary-from-the-process-to-the-modelling

[2] Rtimi R, Sottolichio A, Tassi P. Hydrodynamics of a hyper-tidal estuary influenced by the world's second largest tidal power station (Rance estuary, France). Estuarine, Coastal and Shelf Science. 2021;**250**:107143. DOI: 10.1016/j.ecss.2020.107143

[3] Thiébot J, Guillou S, Brun-Cottan JC. An optimisation method for determining permeability and effective stress relationships of consolidating cohesive sediment deposits. Continental Shelf Research. 2011;**31**(10) Supplement 1: S117-S123. DOI: 10.1016/j.csr.2010.12.004

[4] Hsu T, Jenkins JT, Liu PLF. On two-phase sediment transport: Dilute flow. Journal of Geophysical Research. 2003; **108**(C3):3057. DOI: 10.1029/2001JC001276

[5] Amoudry L, Hsu TJ, Liu PLF. Two-phase model for sand transport in sheet flow regime. Journal of Geophysical Research. 2008;**113**:C03011. DOI: 10.1029/2007JC004179

[6] Nguyen KD, Guillou S, Chauchat J, Barbry N. A two-phase numerical model for suspended-sediment transport in estuaries. Advances in Water Resources. 2009;**32**:1187-1196. DOI: 10.1016/j.advwatres.2009.04.001

[7] Chauchat J, Guillou S. On turbulence closures for two-phase sediment-laden flow models. Journal of Geophysical Research. 2008;**113**:C11017. DOI: 10.1029/2007JC004708

[8] Chauchat J, Médale M. A 3D numerical model for incompressible two-phase flow of a granular bed submitted to a laminar shearing flow. Computer Methods in Applied Mechanics and Engineering. 2010;**199**:439-449. DOI: 10.1016/j.cma.2009.07.007

[9] Nguyen DH, Levy F, Pham Van Bang D, Guillou S, Nguyen KD, Chauchat J. Simulation of dredged sediment releases into homogeneous water using a two-phase model. Advances in Water Resources. 2012;**48**:102-112. DOI: 10.1016/j.advwatres.2012.03.009

[10] Khaled F, Guillou SS, Méar Y, Ferhat H. Impact of blockage ratio on the transport of sediments in the presence of a hydrokinetic turbine: Numerical modelling of the interaction sediments-turbine. International Journal of Sediment Research. 2021;**36**:696-710. DOI: 10.1016/j.ijsrc.2021.02.003

[11] Glowinski R, Pan T-W, Hesla TI, Joseph DD. A distributed lagrange multiplier/fictitious domain for particulate flows. The International Journal of Multiphase Flow. 1999;**25**: 755-794. DOI: 10.1016/S0301-9322(98)00048-2

[12] Peskin CS. The immersed boundary method. Acta Numerica. 2002;**11**:479-517. DOI: 10.1017/S0962492902000077

[13] Yu Z, Shao X. A direct-forcing fictitious domain method for particulate flows. Journal of Computational Physics. 2007;**227**:292-314. DOI: 10.1016/j.jcp.2007.07.027

[14] Wachs A. A DEM-DLM/FD method for direct numerical simulation of particulate flows: Sedimentation of polygonal isometric particles in a Newtonian fluid with collisions. Computers & Fluids. 2010;**38**(8): 1608-1628. DOI: 10.1016/j.compfluid.2009.01.005

[15] Davis RH. In: Tory EM, editor. Velocities of Sedimenting Particles in Suspension, in: Sedimentation of Small Particles in Viscous Fluid, Advances in Fluid Mechanics. Southampton: Computational Mechanics Publications; 1996. pp. 161-198

[16] Guillou S, Makhloufi R. Effect of a shear-thickening rheological behaviour on the friction coefficient in a plane channel flow: A study by direct numerical simulation. Journal for Non-Newtonian Fluid Mechanics. 2007;**144**: 73-86. DOI: 10.1016/J.JNNFM.2007.03.008

[17] Verjus R, Guillou SS, Ezersky A, Angilella JR. Chaotic sedimentation of particle pairs in a vertical channel at low Reynolds number: Multiple states and routes to chaos. Physics of Fluids. 2016;**28**:123303. DOI: 10.1063/1.4968559

[18] Verjus R. Contribution à la modélisation des processus de sédimentation-consolidation dans les ports et les estuaires: Etude bi-échelle. [thesis]. University of Nancy; 2015

[19] Yu Z, Phan-Thien N, Fan Y, Tanner RI. Viscoelastic mobility problem of a system of particles. Journal for Non-Newtonian Fluid Mechanics. 2002;**104**: 87-124. DOI: 10.1016/S0377-0257(02)00014-9

[20] Happel J, Brenner H. Low Reynolds Number Hydrodynamics, 552P. The Hague: Martinus Nijhof; 1973

[21] Jayaweera KOLF, Mason BJ. The behaviour of freely falling cylinders and cones in a viscous fluid. Journal of Fluid Mechanics. 1965;**22**:709-720. DOI: 10.1017/S002211206500109X

[22] Feng J, Hu HH, Joseph DD. Direct simulation of initial-value problems for the motion of solid bodies in a Newtonian fluid. 1. Sedimentation. Journal of Fluid Mechanics. 1994;**261**: 95-134. DOI: 10.1017/S0022112094000285

[23] Richardson JF, Zaki WN. Sedimentation and fluidization: Part I. Transactions of the Institution of Chemical Engineers. 1954;**32**:35-53. DOI: 10.1016/S0263-8762(97)80006-8

[24] Guazzelli E, Morris JF. A Physical Introduction to Suspension Dynamics. Cambridge: Cambridge University Press; 2012

Chapter 5

Efficient Simulation Tools (EST) for Sediment-Laden Shallow Flows

Sergio Martínez-Aranda and Pilar García-Navarro

Abstract

Rapid flows of water-sediment mixtures are probably the most challenging and unknown geophysical gravity-driven processes. The fluidized material in motion consists of a mixture of water and multiple solid phases with different specific characteristics. Modeling sediment transport involves an increasing complexity due to the variable bulk properties in the sediment-water mixture, the coupling of physical processes, and the presence of multiple layer phenomena. Two-dimensional shallow-type mathematical models are built in the context of free surface flows and are applicable to most of these geophysical surface processes. Their numerical solution in the finite volume framework is governed by the dynamical properties of the equations, the coupling between flow variables and the computational grid. The complexity of the numerical resolution of these highly unsteady flows and the computational cost of simulation tools increase considerably with the refinement of the non-structured spatial discretization, so that the computational effort required is one of the biggest challenges for the application of depth-averaged 2D models to large-scale long-term flows. Throughout this chapter, the combination of 2D mathematical models, robust numerical methods, and efficient computing kernels is addressed to develop Efficient Simulation Tools (EST's) for environmental surface processes involving sediment transport with realistic temporal and spatial scales.

Keywords: sediment-laden flows, depth-averaged rheological models, bed-material entrainment, shallow flow models, GPU high-performance-computing

1. Introduction

Sediment transport is ubiquitous in environmental water bodies such as rivers, floods, coasts and estuaries, but also is the main process in natural ladslides, debris flows, muddy slurries or mining tailings (**Figure 1**). These are considered highly solid-laden fluids, where the density of the water-solid mixture can be more than twice or three times the water density and the bulk solid phase represents about 40–80% of the flow volume [1].

The presence of the solid phases, especially the fine material as silt or clay, affects the rheological behavior of the mixture. The water-sediment mixture rheology begins

Figure 1.
Geophysical surface flows involving sediment transport.

to be affected by fine solid particle transported in the flow when the volumetric concentration of fine sediment particles reaches about 4% by volume [2], creating a slight shear strength within the fluid. For higher concentrations, the mixture shows a marked non-Newtonian rheology. Mud and debris flows lie between hyperconcentrated flows and wet avalanches [3]. High concentrations of solids generate a critical yield stress which allows that gravel particles can be suspended indefinitely into the fluidized material.

The mass exchange between mud/debris mixtures and erodible beds involves complicated physical processes and the understanding of its theoretical basis remains unclear. Experiments in large-scale channel [4, 5] and field observations in real debris events [6, 7] indicate that the entrainment volume in steep beds can be in the same order of magnitude as the initial volume mobilized. Debris and mud flows gain much of their mass and momentum as they move over steep slopes as a consequence of the material entrainment from the erodible bed, before deposition begins on flatter terrain downstream.

The mathematical modeling of solid-liquid mixture flows and their numerical resolution is still a challenging topic, especially when dealing with realistic applications. When liquid and solid phases are well-mixed, assuming that the solid phase is distributed uniformly over the flow column allows the use of depth-averaged models derived from the vertical integration of the Navier-Stokes equations [8]. Shallow-type mathematical models represent a simplified formulation, derived from the general 3D Navier-Stokes equations, which is applicable to a large number of these geophysical surface processes involving sediment transport. The simplest models, used in river and coastal dynamics, assume small enough sediment concentrations throughout the flow to consider the bulk density constant and uniform. Most of the numerical models reported for highly solid-laden flows also use this one-single-phase approach, neglecting the bulk density in the shallow-flow mass and momentum equations [9, 10]. Nevertheless, even small density gradients influence importantly the mixing dynamics in flow confluences [11] and larger gradients can also generate numerical oscillations and instabilities throughout mixing interfaces.

Modeling sediment transport involves an increasing complexity with respect to rigid-bed shallow water models [12] due to the presence of variable sediment-fluid mixture properties, coupling of physical processes and multiple layers phenomena [13]. One of the biggest challenges for the application of depth-averaged models to realistic large-scale long-term flows is the computational effort required. New strategies to reduce the computational effort have been developed in the last decade through the use of parallelization techniques based on Multiprocessing (OpenMP) or Message Passing Interface (MPI), which allow to run simulations on multi-CPU clusters. Their main drawback is the associated hardware cost and energy requirements, which are directly proportional to the number of CPU-cores available and limit their efficiency. In the last years, the usage of Graphics Processing Units (GPU) hardware accelerators for sequential computation has demonstrated to be an efficient and low cost alternative to the traditional multi-CPU strategies [14]. GPU-accelerated algorithms have been developed for real-time floods forecasting [15], real-scale bedload erosive shallow-flows [16] or tsunami prediction [17]. GPU devices are oriented to perform arithmetical operations on vector-based information. Unlike the conventional shared-memory multi-CPU implementations, the GPU solution must be designed taking into account the fact that the GPU is an independent device with its own RAM memory. This means that the memory transfer between the conventional RAM memory and the GPU device memory plays a key role in the performance of GPU-accelerated software.

2. Depth-integrated equations for shallow flows

2.1 Mass and linear momentum conservation

The flow of a water-sediment mixture can be mathematically described assuming the movement of the solid particles as a diffusion phenomenon into the liquid phase. Then, the continuity and momentum conservation for the mixture, supplemented with the transport equation for the solid phase, can be established for modeling these two-phase flows. Although both solid and liquid phases are incompressible when considered independently, the bulk behavior of the solid-liquid mixture is the same as that of a compressible material depending on the local solid phase volumetric concentration. It possible to define the bulk density $\rho = \rho_w n + \rho_s \phi$ and linear momentum $\rho \mathbf{u} = \rho_w n \mathbf{u_w} + \rho_s \phi \mathbf{u_s}$ of the mixture, being ρ_w and ρ_s the density of the fluid and solid phases, respectively, ϕ the volumetric solid-phase concentration and $n = 1 - \phi$ the volumetric fluid-phase fraction or mixture porosity, $\mathbf{u_w}$ the velocity of the pore-fluid and $\mathbf{u_s}$ the advective sediment particle velocity.

The 3D time-averaged Navier-Stokes equations for mass and momentum conservation of a two-phase mixture can be written in the Cartesian coordinate system $\mathbf{X} = (x, y, z)$ as

$$\frac{\partial \rho}{\partial t} + \nabla \cdot (\rho \mathbf{u}) = 0 \tag{1}$$

$$\frac{\partial (\rho \mathbf{u})}{\partial t} + \nabla \cdot (\rho \mathbf{u} \otimes \mathbf{u}) = \mathbf{F} - \nabla p + \nabla \cdot \tau \tag{2}$$

where $\mathbf{u} = (u_x, u_y, u_z)$ is the bulk velocity in any point of the fluidized material, $\mathbf{F} = (F_x, F_y, F_z)$ are the external forces, such as gravity and p denotes the pressure of the mixture. The term $\tau = \tau_{ij}\ \ (i, j = x, y, z)$ is the deviatoric stress tensor. With low solid-phase concentrations, the water-sediment mixture behaves as a Newtonian fluid, with a constitutive relation given by the Navier-Poisson law [8]. Nevertheless, for high sediment concentrations the mixture becomes a kind of non-Newtonian fluid with a complex constitutive law, which depends on multiple factors, relating stresses and deformation rates.

In order to develop a shallow-type depth-averaged mathematical model, the Navier-Stokes system (1) and (2) is integrated between the free surface $z_s = z_s(t, x, y)$ and the bottom surface of the flow column $z_b = z_b(t, x, y)$, which is also considered a movable interface. The kinematic conditions at these boundaries can be expressed as

$$\frac{\partial z_s}{\partial t} + (u_x)_s \frac{\partial z_s}{\partial x} + (u_y)_s \frac{\partial z_s}{\partial y} = (u_z)_s \tag{3}$$

$$\frac{\partial z_b}{\partial t} + (u_x)_b \frac{\partial z_b}{\partial x} + (u_y)_b \frac{\partial z_b}{\partial y} = (u_z)_b + N_b \tag{4}$$

being N_b the net volumetric flux through the bed interface along the $z-$coordinate, being positive for net entrainment conditions and negative for net deposition fluxes. The subscripts $(\cdot)_s$ and $(\cdot)_b$ indicate the value of the corresponding variable at the flow free surface and the bottom bed interface respectively.

The mass conservation Eq. (1) is integrated along the water column as

$$\int_{z_b}^{z_s} \partial_t(\rho)\,\mathrm{d}z + \int_{z_b}^{z_s} \partial_x(u_x\rho)\,\mathrm{d}z + \int_{z_b}^{z_s} \partial_y(u_y\rho)\,\mathrm{d}z + \int_{z_b}^{z_s} \partial_z(u_z\rho)\,\mathrm{d}z = 0 \tag{5}$$

Applying Leibnitz's rule to each term of (5) and using (3) and (4) leads to

$$\frac{\partial(\overline{\rho}h)}{\partial t} + \frac{\partial}{\partial x}(\overline{\rho}h\overline{u}) + \frac{\partial}{\partial y}(\overline{\rho}h\overline{v}) = -(\rho)_b N_b \tag{6}$$

where the depth-averaged of the mixture bulk density $\overline{\rho}$ and velocities are defined as

$$\overline{\rho}h = \int_{z_b}^{z_s} \rho\,\mathrm{d}z \qquad \overline{u} = \frac{1}{\overline{\rho}h}\int_{z_b}^{z_s} u_x\rho\,\mathrm{d}z \qquad \overline{v} = \frac{1}{\overline{\rho}h}\int_{z_b}^{z_s} u_y\rho\,\mathrm{d}z \tag{7}$$

being $h = z_s - z_b$ the flow depth and $(\rho)_b$ the mixture bulk density at the bed interface z_b.

Regarding the momentum depth integration, the volumetric force components in (1) and (2) are considered for the sake of simplicity as $F_x = F_y = 0$ and $F_z = -\rho g$. Moreover, the z-momentum equation in (2) can be simplified by assuming shallow-flow scaling and neglecting temporal, convective and stress terms, leading to a hydrostatic pressure distribution along the flow column.

The x-momentum in (2) is integrated throughout the flow depth in the following way

$$\begin{aligned}\int_{z_b}^{z_s} \partial_t(\rho u_x)\,\mathrm{d}z + \int_{z_b}^{z_s} \partial_x(u_x\rho u_x)\,\mathrm{d}z + \int_{z_b}^{z_s} \partial_y(u_y\rho u_x)\,\mathrm{d}z + \int_{z_b}^{z_s} \partial_z(u_z\rho u_x)\,\mathrm{d}z \\ = -\int_{z_b}^{z_s} \partial_x(p)\,\mathrm{d}z + \int_{z_b}^{z_s} \left(\partial_x\tau_{xx} + \partial_y\tau_{xy} + \partial_z\tau_{xz}\right)\mathrm{d}z\end{aligned} \tag{8}$$

Applying Leibnitz's rule and considering the kinematic boundary conditions at both the free surface (3) and the bed interface (4), the left hand side of (8) is integrated as

$$\frac{\partial(\overline{\rho}h\overline{u})}{\partial t} + \frac{\partial}{\partial x}\left(\overline{\rho}h\overline{u}^2 - \overline{\rho}hD_{xx}\right) + \frac{\partial}{\partial y}\left(\overline{\rho}h\overline{u}\overline{v} - \overline{\rho}hD_{xy}\right) + (\rho u_x)_b N_b \tag{9}$$

with (D_{xx}, D_{xy}) accounting for the depth-averaged dispersion momentum transport due to the non-uniformity of the vertical velocity profile, and defined as

$$D_{xx} = -\frac{1}{\overline{\rho}h}\int_{z_b}^{z_s} \rho(u_x - \overline{u})^2\,\mathrm{d}z \tag{10}$$

$$D_{xy} = -\frac{1}{\overline{\rho}h}\int_{z_b}^{z_s} \rho(u_x - \overline{u})(u_y - \overline{v})\,\mathrm{d}z \tag{11}$$

On the right hand side of (8), using the hydrostatic pressure distribution and assuming that the vertical density gradient is negligible compared with those along the horizontal plane, as occurs in natural debris and mud slurry flows, the integral of the pressure gradient along the $x-$coordinate can be expressed as

$$-\int_{z_b}^{z_s} \partial_x(p)\,\mathrm{d}z = -g\int_{z_b}^{z_s}\left(\frac{\partial}{\partial x}\int_{z}^{z_s}\rho\,\mathrm{d}z\right)\mathrm{d}z = \underbrace{-\frac{\partial}{\partial x}\left(\frac{1}{2}g\overline{\rho}h^2\right)}_{\text{Conservative term}} \quad \underbrace{-g\overline{\rho}h\frac{\partial z_b}{\partial x}}_{\text{Bed}-\text{pressure term}} \tag{12}$$

separating the pressure gradient term into a conservative component plus a bed-pressure component.

The stress terms on the right hand side of (8) are also integrated along the flow column, leading to the final expresion for the depth-integrated momentum equation along the $x-$coordinate

$$\begin{aligned}\frac{\partial(\overline{\rho}h\overline{u})}{\partial t} + \frac{\partial}{\partial x}\left(\overline{\rho}h\overline{u}^2 + \frac{1}{2}g\overline{\rho}h^2\right) + \frac{\partial}{\partial y}\left(\overline{\rho}h\overline{u}\overline{v}\right) = -g\overline{\rho}h\frac{\partial z_b}{\partial x} + \tau_{sx} - \tau_{bx} \\ + \frac{\partial}{\partial x}\left(\overline{\rho}h(T_{xx} + D_{xx})\right) + \frac{\partial}{\partial y}\left(\overline{\rho}h(T_{xy} + D_{xy})\right) \\ -(\rho u_x)_b N_b\end{aligned} \tag{13}$$

and, similarly, for the $y-$coordinate. The boundary stress terms in (13) at the free surface z_s and the bed interface z_b are

$$\tau_{sx} = \left(\tau_{xz}\right)_s - \left(\tau_{xx}\right)_s \frac{\partial z_s}{\partial x} - \left(\tau_{xy}\right)_s \frac{\partial z_s}{\partial y} \tag{14}$$

$$-\tau_{bx} = -\left(\tau_{xz}\right)_b + \left(\tau_{xx}\right)_b \frac{\partial z_b}{\partial x} + \left(\tau_{xy}\right)_b \frac{\partial z_b}{\partial x} \tag{15}$$

where τ_{sx} denotes the $x-$coordinate component of the wind action at the free surface, whereas τ_{bx} represents the $x-$coordinate component of the boundary shear stress at the bed interface, opposing to the flow movement. Furthermore, the depth-integrated stress terms along the flow column $\left(T_{xx}, T_{xy}\right)$ are defined as

$$T_{xx} = \frac{1}{\overline{\rho} h} \int_{z_b}^{z_s} \tau_{xx}\, dz \quad T_{xy} = \frac{1}{\overline{\rho} h} \int_{z_b}^{z_s} \tau_{xy}\, dz \tag{16}$$

2.2 Depth-integrated solid transport equation

The solid-phase transport process is governed by

$$\frac{\partial(\rho_s \phi)}{\partial t} + \nabla \cdot (\rho_s \phi \mathbf{u_s}) = 0 \tag{17}$$

where $\mathbf{u_s} = \left(u_{sx}, u_{sy}, u_{sz}\right)$ is the local velocity of the solid particles. Because the solid-phase velocity is not included in the dependent variables of the fluid dynamic system, and assuming that the sediment particles are incompressible and non-porous, (17) is rewritten as

$$\frac{\partial \phi}{\partial t} + \nabla \cdot (\phi \mathbf{u}) = \nabla \cdot [(\mathbf{u} - \mathbf{u_s})\phi] \tag{18}$$

where the term on the right hand side accounts for the drag effects caused by the liquid phase on the advective solid flux. Due to the definition of the bulk linear momentum of the mixture, this term is usually neglected and the approximation $\mathbf{u} \approx \mathbf{u_s}$ is actually assumed.

The transport equation (18) must also be integrated throughout the entire flow column defining the total solid phase being transported in the flow column as

$$\overline{\phi} h = \int_{z_b}^{z_s} \phi\, dz \tag{19}$$

where $\overline{\phi}$ is the dept-averaged volumetric solid concentration. Therefore, applying the Leibnitz's integration rule to (18) and considering the kinematic boundary conditions at both the free surface (3) and the bed interface (4), the depth-integrated equation for the solid phase in the flow column reduces to

$$\frac{\partial(\overline{\phi} h)}{\partial t} + \frac{\partial}{\partial x}\left(\overline{\phi} h \overline{u}\right) + \frac{\partial}{\partial y}\left(\overline{\phi} h \overline{v}\right) = -(\phi)_b N_b + \frac{\partial}{\partial x}\left(\overline{\phi} h D_{sx}\right) + \frac{\partial}{\partial y}\left(\overline{\phi} h D_{sy}\right) \tag{20}$$

where $(\phi)_b$ denotes the solid concentration in the bed layer surface z_b and $\left(D_{sx}, D_{sy}\right)$ account for the depth-averaged dispersive flux due to the non-uniformity of both the velocity and solid concentration profiles throughout the flow column, defined as

$$D_{sx} = -\frac{1}{\overline{\phi} h} \int_{z_b}^{z_s} \rho (u_x - \overline{u}) \left(\frac{\phi}{\rho} - \frac{\overline{\phi}}{\overline{\rho}} \right) \mathrm{d}z \tag{21}$$

$$D_{sy} = -\frac{1}{\overline{\phi} h} \int_{z_b}^{z_s} \rho (u_y - \overline{v}) \left(\frac{\phi}{\rho} - \frac{\overline{\phi}}{\overline{\rho}} \right) \mathrm{d}z \tag{22}$$

Furthermore, assuming that the volumetric solid concentration in the bed layer is $1 - \xi$ with ξ denoting the bulk porosity of the bed layer, the solid mass conservation in the bed layer requires that

$$(1 - \xi) \frac{\partial z_b}{\partial t} = (\phi)_b N_b \tag{23}$$

2.3 Constitutive model for complex viscoplastic flows

So far, there is not a universal closure relation for representing the viscous terms in complex non-Newtonian flows. Constitutive formulations used for environmental sediment-water mixtures are mainly derived from 3D general rheological models which, assuming isotropic material and isochoric flow, allow to express the deviatoric component of the stress tensor σ in the material as

$$\tau = \Phi_1(I_{2D}) \mathbf{D} \tag{24}$$

where $\mathbf{D} \equiv D_{ij} = \frac{1}{2}\left(\partial_j u_i + \partial_i u_j\right) \ (i, j = x, y, z)$ is the rate of deformation tensor and Φ_1 is a scalar function of the second invariant $I_{2D} = \frac{1}{2}\mathrm{tr}\left(\mathbf{D}^2\right)$ of the rate of deformation tensor $\mathbf{D}$ [18] and depends on multiple factor, such as cohesive stress, pore-fluid pressure or flow initial regime. The generalized viscoplastic model, also called Herschel-Bulkley model, assumes dependence of Φ_1 on the three parameters: τ_0 the cohesive-frictional yield strength, K a consistency coefficient, and m a parameter characterizing the rheological response of the mixture [13, 18], known as behavior index. It is worth mentioning that the dimensions of consistency coefficient K depends on the behavior index m. Therefore, the function Φ_1 is expressed as

$$\Phi_1(I_{2D}) = \frac{\tau_0}{\sqrt{I_{2D}}} + 2K\left(4I_{2D}\right)^{\frac{m-1}{2}} \tag{25}$$

Considering simple shear state along the flow direction, the velocity vector throughout the flow column is expressed as $\mathbf{u} = \left(U(z)n_{ux},\ \ U(z)n_{uy},\ \ 0\right)$, where $U(z)$ is the modulus of the bulk mixture velocity and $\mathbf{n_u} = \left(n_{ux}, n_{uy}\right)$ is the velocity unit vector. Therefore

$$\mathbf{D} = \begin{pmatrix} 0 & 0 & \frac{1}{2}\frac{\mathrm{d}U}{\mathrm{d}z} n_{ux} \\ 0 & 0 & \frac{1}{2}\frac{\mathrm{d}U}{\mathrm{d}z} n_{uy} \\ \frac{1}{2}\frac{\mathrm{d}U}{\mathrm{d}z} n_{ux} & \frac{1}{2}\frac{\mathrm{d}U}{\mathrm{d}z} n_{uy} & 0 \end{pmatrix} \quad I_{2D} = \frac{1}{4}\left(\frac{\mathrm{d}U}{\mathrm{d}z}\right)^2 \tag{26}$$

Replacing (26) and (25) into (24) leads to

$$\tau = \begin{pmatrix} 0 & 0 & \tau(z)n_{ux} \\ 0 & 0 & \tau(z)n_{uy} \\ \tau(z)n_{ux} & \tau(z)n_{uy} & 0 \end{pmatrix} \tag{27}$$

being $\tau(z)$ the shear stress along the flow direction, which depends on the fluid rheology function (25). For the Newtonian constitutive model, the yield strength τ_0 is null, $K \ [Pa \cdot s]$ is the dynamic viscosity of the fluid μ and $m = 1$. The generalized model reduces to $\tau(z) = \mu \frac{dU}{dz}$ for viscous Newtonian fluids. A widespread non-Newtonian constitutive relation for geophysical surface flows in laminar regime is the Bingham model, which considers a pure cohesive yield stress $\tau_0 = \tau_y$ for the flow initiation, $K = \mu_B \ [Pa \cdot s]$ the Bingham plastic viscosity and $m = 1$. In this case, the generalized model reduces to $\tau(z) = \tau_y + \mu_B \frac{dU}{dz}$.

The frictional non-linear viscoplastic model considers a Coulomb-Terzaghi linear relation between the effective normal stress $\sigma_e(z)$ and the shear stress, hence

$$\tau_0 = \sigma_e(z) \tan\delta_f = [\rho g(z_s - z) - \mathcal{P}(z)] \tan\delta_f \tag{28}$$

where $\mathcal{P}(z)$ denotes de pore-fluid pressure and δ_f accounts for the effective friction angle bewteen solid particles [19]. Estimation of the pore pressure distribution $\mathcal{P}(z)$ is a challenging task [1, 19, 20], although its effects on the reduction of the intergranular shear stress seem to be demonstrated [5, 21, 22]. The simplest models divide the pore pressure into a hydrostatic component plus a dynamic additive component $\mathcal{P}(z) = (1 + \mathcal{E})\rho_w g(z_s - z)$, being ρ_w the density of the pore-liquid and $\mathcal{E}$ a tunning coefficient which usually takes values from about 0.4 to 0.8.

Using (28), and considering a plastic viscosity $K = \mu_P \ [Pa \cdot s^m]$, the generalized model reduces to

$$\tau(z) = \sigma_e(z) \tan\delta_f + \mu_P \left(\frac{dU}{dz}\right)^m \tag{29}$$

Formulation of closure models for the bottom shear stress requires to integrate the deviatoric stress tensor (24) throughout the flow column. This is not a trivial problem since the structure of the flow along the vertical direction is lost and only the averaged quantities are available. Assuming the deviatoric stress tensor (27), all the depth-averaged stress terms (T_{xx}, T_{xy}) and (T_{yx}, T_{yy}) (16) are null, and

$$\tau_{bx} = (\tau_{xz})_b - (\tau_{xx})_b \frac{\partial z_b}{\partial x} - (\tau_{xy})_b \frac{\partial z_b}{\partial y} = \tau_b n_{ux} \tag{30}$$

$$\tau_{by} = (\tau_{yz})_b - (\tau_{yx})_b \frac{\partial z_b}{\partial x} - (\tau_{yy})_b \frac{\partial z_b}{\partial y} = \tau_b n_{uy} \tag{31}$$

where $\tau_b = \tau(z_b)$ is the shear stress at the basal interface along the flow direction. In order to obtain a depth-averaged formulation for the basal shear stress τ_b, the distributed shear stress function $\tau(z)$ must be integrated along the flow column. This allows to obtain both the velocity distribution along the vertical direction and the basal resistance τ_b expressed as a function of the depth-averaged flow variables. If the pore-pressure excess in (28) is considered linear, the the constitutive eq. (29) can be rewritten as

$$\tau(z) = \tau_f\left(1 - \frac{z - z_b}{h}\right) + \mu_P\left(\frac{\mathrm{d}U}{\mathrm{d}z}\right)^m \tag{32}$$

being $\tau_f = (\rho g h - \mathcal{P}_b)\tan\delta_f$ the value of the frictional yield stress at the basal surface and $\mathcal{P}_b$ is the pore-pressure at the bed surface (**Figure 2**).

Therefore, assuming the induced shear distribution along the flow column follows

$$\tau(z) = \tau_b\left(1 - \frac{z - z_b}{h}\right) \tag{33}$$

the velocity derivative along the vertical direction can expressed as

$$\frac{\mathrm{d}U}{\mathrm{d}z} = \left[\frac{\tau_b - \tau_f}{\mu_P}\left(1 - \frac{z - z_b}{h}\right)\right]^{1/m} \tag{34}$$

and integrating (34) leads to the velocity vertical distribution

$$U(z) = \frac{m}{m+1}\left(\frac{\tau_b - \tau_f}{\mu_P}\right)^{1/m} h\left[1 - \left(1 - \frac{z - z_b}{h}\right)^{\frac{m+1}{m}}\right] \tag{35}$$

for the non-linear viscoplastic model. Note that the velocity at the free surface $U(z_s) \equiv U_h$ and the depth averaged velocity $\overline{U}$ can be expressed as

$$U_h = \frac{m}{m+1}\left(\frac{\tau_b - \tau_f}{\mu_P}\right)^{1/m} h \qquad \overline{U} = \frac{m+1}{2m+1} U_h \tag{36}$$

Integrating (35) throughout the flow column allows to obtain the basal shear stress τ_b as a function of the flow depth h, the averaged flow velocity $\overline{U} = \sqrt{\overline{u}^2 + \overline{v}^2}$ and the basal frictional yield strength τ_f using

$$\tau_b = \tau_f + \left(\frac{2m+1}{m}\right)^m \mu_P\left(\frac{\sqrt{\overline{u}^2 + \overline{v}^2}}{h}\right)^m \tag{37}$$

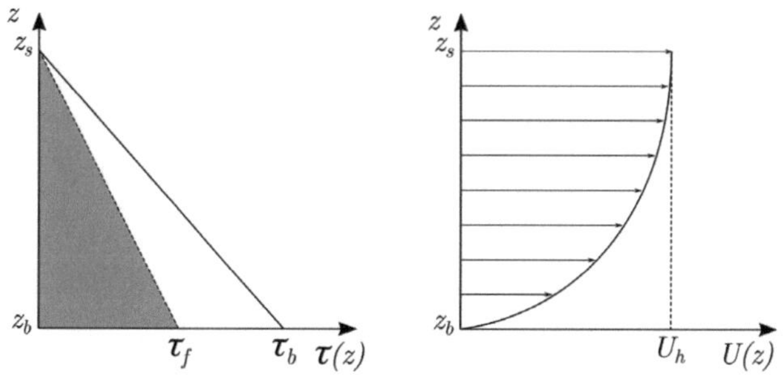

Figure 2.
Velocity and stress distribution for the generalized non-linear frictional model.

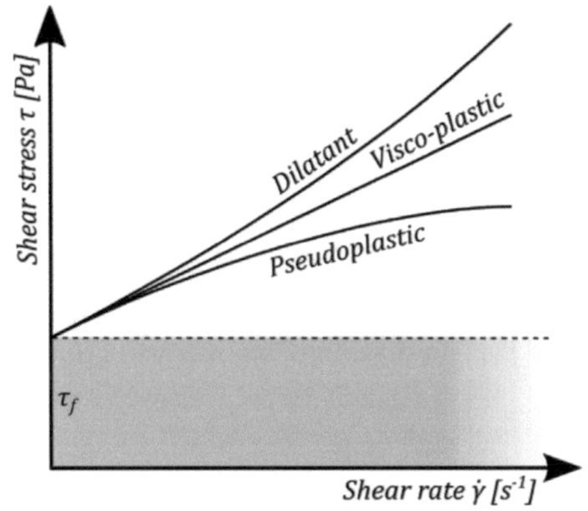

Figure 3.
Basal resistance behavior for the generalized frictional non-linear viscoplastic model.

It is worth mentioning that (37) represents a generalized depth-integrated formulation for viscoplastic flows (**Figure 3**) which encompasses: pseudoplastic or shear-thinning behavior for $m < 1$, reducing the apparent viscosity as the induced shear rate increases; linear viscoplastic behavior for $m = 1$, with a linear relation between shear stress and shear rate; and dilatant or shear-thickening behavior for $m > 1$, increasing the apparent viscosity as the induced shear rate grows.

2.4 Net mass exchange between bed and flow layers

The net volumetric exchange N_b between the underlying bed layer and the mixture flow column at the bed surface $z_b(t, x, y)$, appearing in (6), (20) and (23), is usually modeled as the balance between the entrainment and the deposition vertical solid fluxes, E_b and D_b respectively, leading to

$$N_b = \frac{1}{(\phi)_b}(D_b - E_b) \tag{38}$$

involving the bed surface solid concentration $(\phi)_b = 1 - \xi$.

The deposition rate D_b is commonly related to the solid particles settling velocity in the mixture ω_{sm} and to the near-bed solid concentration in the flow $\phi_{z\to z_b}$, which is usually related to the depth-averaged solid concentration in the flow as $\phi_{z\to z_b} = \alpha\overline{\phi}$, being α an adaptation or recovery coefficient, leading to $D_b = \alpha\omega_{sm}\overline{\phi}$.

The settling velocity of the solid particles ω_{sm} in highly concentrated mixtures is influenced by the presence of other solid particles. Furthermore, in dense-packed mixtures with moderate plastic fine fractions in the flow column such as muddy slurries, the particle settling velocity can be strongly reduced by the development of internal yield stresses in the pore-fluid. Richardson and Zaki [23] proposed $\omega_{sm} = \left(1 - \overline{\phi}\right)^m \omega_s$, being ω_s the settling velocity of the sediment particles in clear water and m a hindering empirical exponent depending on the Reynolds particle number ($Re_p = \omega_s d_s/\nu$, with ν the clear water kinematic viscosity) which usually takes values close to $m = 4$. Therefore

$$D_b = \alpha_D \omega_s \overline{\phi} \left(1 - \overline{\phi}\right)^4 \tag{39}$$

where α_D is a dimensionless parameter which accounts for mulple factor and requires calibration.

The erosion solid flux E_b is directly related to the turbulent fluctuation of the volumetric solid concentration and flow velocity near the bed surface. We assume that this near-bed erosion rate is at the capacity of the flow to entrain solid material from the underlying bed layer, hence it is related to the settling velocity of the particles in clear water ω_s and the near equilibrium concentration $\phi^*_{z \to z_b}$. The near-bed equilibrium concentration is related to the depth-averaged equilibrium concentration $\overline{\phi^*}$ as $\phi^*_{z \to z_b} = \alpha^* \overline{\phi^*}$, being α^* an adaptation coefficient under equilibrium conditions. When equilibrium solid transport states are reached, the adaptation coefficients α and α^* coincide but in non-equilibrium states $\alpha^* \neq \alpha$ generally. Therefore, for the sake of simplicity, we assume that the vertical erosion rate E_b is expressed as

$$E_b = \alpha_E \omega_s \overline{\phi^*} \tag{40}$$

where α_E is a dimensionless empirical parameter which requires calibration. The capacity solid concentration is usually computed as $\overline{\phi^*} = q_s^* / \left(h\sqrt{\overline{u}^2 + \overline{v}^2}\right)$, where q_s^* accounts for the value of the solid transport throughout the flow column in capacity or equilibrium condition, which can be estimated using the multiple empirical relationships from the local hydrodynamic variables [8].

3. Efficient numerical tools for multi-grain mud/debris flows

Considering a multi-grain mixture flow, the resulting system is composed by $3 + N + 1$ conservation equations accounting for the bulk mass (6) and momentum (13), the transport of the N sediment classes (20) and the bed elevation evolution (23). The dimensionless bulk density r can be expressed by defining a new variable ϕ^χ, referred to as buoyant solid concentration

$$r = \frac{\rho}{\rho_w} = 1 + \phi^\chi \quad \text{with :} \quad \phi^\chi = \sum_{p=1}^{N} \frac{\rho_{s,p} - \rho_w}{\rho_w} \phi_p \tag{41}$$

where $\rho_{s,p}$ and ϕ_p are the density and depth-averaged volumetric concentration of the pth solid phase respectively. Using (41), the equations forming the system can be recast as five conservation laws and rewritten in vector form as

$$\frac{\partial \mathbf{U}}{\partial t} + \nabla \cdot \mathbf{E}(\mathbf{U}) = \mathbf{S_b}(\mathbf{U}) + \mathbf{S}_\tau(\mathbf{U}) + \mathbf{E_b}(\mathbf{U}) \tag{42}$$

where $\mathbf{U}$ is the vector of conservative variables and $\mathbf{E}(\mathbf{U}) = (\mathbf{F}(\mathbf{U}), \mathbf{G}(\mathbf{U}))$ are the convective fluxes along the $\mathbf{X} = (x, y)$ horizontal coordinates, respectively, expressed as

$$\mathbf{U} = \begin{pmatrix} rh \\ rhu \\ rhv \\ h\phi^\chi \\ z_b \end{pmatrix} \quad \mathbf{F}(\mathbf{U}) = \begin{pmatrix} rhu \\ rhu^2 + \frac{1}{2}grh^2 \\ rhuv \\ hu\,\phi^\chi \\ 0 \end{pmatrix} \quad \mathbf{G}(\mathbf{U}) = \begin{pmatrix} rhv \\ rhuv \\ rhv^2 + \frac{1}{2}grh^2 \\ hv\,\phi^\chi \\ 0 \end{pmatrix} \tag{43}$$

The vector $\mathbf{S_b}(\mathbf{U})$ accounts for the momentum source term associated to the variation of the pressure force on the bed interface, whereas $\mathbf{S}_\tau(\mathbf{U})$ is the momentum dissipation due to the basal resistance. Finally, the source term $\mathbf{E_b}(\mathbf{U})$ accounts for the bulk mass exchange between the mixture flow and the bed layer.

$$\mathbf{S_b}(\mathbf{U}) = \begin{pmatrix} 0 \\ -grh\dfrac{\partial z_b}{\partial x} \\ -grh\dfrac{\partial z_b}{\partial y} \\ 0 \\ 0 \end{pmatrix} \quad \mathbf{S}_\tau(\mathbf{U}) = \begin{pmatrix} 0 \\ -\dfrac{\tau_b}{\rho_w}n_{ux} \\ -\dfrac{\tau_b}{\rho_w}n_{uy} \\ 0 \\ 0 \end{pmatrix} \quad \mathbf{E_b}(\mathbf{U}) = \begin{pmatrix} -\dfrac{\rho_b}{\rho_w(1-\xi)}\displaystyle\sum_{p=1}^{N}(D_b - E_b)_p \\ 0 \\ 0 \\ -\displaystyle\sum_{p=1}^{N}\dfrac{\rho_{s,p}-\rho_w}{\rho_w}(D_b - E_b)_p \\ \dfrac{1}{1-\xi}\displaystyle\sum_{p=1}^{N}(D_b - E_b)_p \end{pmatrix} \tag{44}$$

3.1 Finite volume method in unstructured meshes

System (42) is time dependent, non linear and contains mass and momentum source terms. Under the hypothesis of dominant advection it can be classified as belonging to the family of hyperbolic systems. In order to obtain a numerical solution, the spatial domain is divided in computational cells using a fixed-in-time mesh and system (42) is integrated in each cell Ω_i. Applying the Gauss theorem leads to

$$\frac{\mathrm{d}}{\mathrm{d}t}\int_{\Omega_i}\mathbf{U}\,\mathrm{d}\Omega + \sum_{k=1}^{NE}(\mathbf{E}\cdot\mathbf{n})_k l_k = \int_{\Omega_i}\mathbf{S_b}(\mathbf{U})\,\mathrm{d}\Omega + \int_{\Omega_i}\mathbf{S}_\tau(\mathbf{U})\,\mathrm{d}\Omega + \int_{\Omega_i}\mathbf{E_b}(\mathbf{U})\,\mathrm{d}\Omega \tag{45}$$

being NE the number of edges for the i cell, $\mathbf{n} = (n_x, n_y)_k$ the outward unit normal vector, l_k the length of the edge and hence $(\mathbf{E}\cdot\mathbf{n})_k$ the value of the normal flux through the kth edge (**Figure 4**).

The left hand side of (42), also called homogeneous part, satisfies the rotation invariant property [24] and hence the conservative normal flux vector can be rewritten as

$$(\mathbf{E}\cdot\mathbf{n})_k = \left[\mathbf{F}(\mathbf{U})n_x + \mathbf{G}(\mathbf{U})n_y\right]_k = \mathbf{R}_k^{-1}\mathbf{F}(\mathbf{R}_k\mathbf{U}) \tag{46}$$

being $\mathbf{R}$ a rotation matrix which projects the global orthogonal framework $\mathbf{X} = (x, y)$ into the local framework $\hat{\mathbf{X}} = (\hat{x}, \hat{y})$ of the kth cell edge (**Figure 4**),

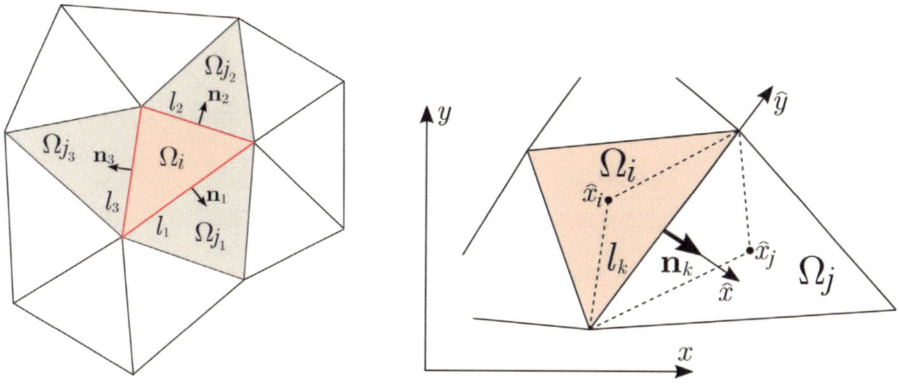

Figure 4.
Computational cells in triangular meshes and local coordinates at the kth cell edge.

corresponding to the normal and the tangential directions to the edge respectively. The rotation matrix $\mathbf{R}_k$ and its inverse $\mathbf{R}_k^{-1}$ are defined as

$$\mathbf{R}_k = \begin{pmatrix} 1 & 0 & 0 & 0 & 0 \\ 0 & n_x & n_y & 0 & 0 \\ 0 & -n_y & n_x & 0 & 0 \\ 0 & 0 & 0 & 1 & 0 \\ 0 & 0 & 0 & 0 & 1 \end{pmatrix}_k \qquad \mathbf{R}_k^{-1} = \begin{pmatrix} 1 & 0 & 0 & 0 & 0 \\ 0 & n_x & -n_y & 0 & 0 \\ 0 & n_y & n_x & 0 & 0 \\ 0 & 0 & 0 & 1 & 0 \\ 0 & 0 & 0 & 0 & 1 \end{pmatrix}_k \tag{47}$$

and the set of local conservative variables $\hat{\mathbf{U}}$ and fluxes $\mathbf{F}(\hat{\mathbf{U}})$ is defined as

$$\hat{\mathbf{U}} \equiv \mathbf{R}_k\mathbf{U} = \begin{pmatrix} rh \\ rh\hat{u} \\ rh\hat{v} \\ h\phi^\chi \\ z_b \end{pmatrix} \qquad \mathbf{F}(\hat{\mathbf{U}})_k \equiv \mathbf{F}(\mathbf{R}_k\mathbf{U}) = \begin{pmatrix} rh\hat{u} \\ rh\hat{u}^2 + \frac{1}{2}g_n rh^2 \\ rh\hat{u}\hat{v} \\ h\hat{u}\,\phi^\chi \\ 0 \end{pmatrix} \tag{48}$$

where $\hat{u} = un_x + vn_y$ and $\hat{v} = -un_y + vn_x$ are the components of the flow velocity $\hat{\mathbf{u}}$ in the local framework.

The value of the fluxes through the kth cell edge can be augmented incorporating the contribution of the momentum source terms $\mathbf{S_b}$ and $\mathbf{S}_\tau$ into the homogeneous normal fluxes $\mathbf{F}(\hat{\mathbf{U}})$ [25]. The momentum terms can be included within the local framework $(\hat{x}, \hat{y})$ using the spatial discretization

$$\int_{\Omega_i} \mathbf{S_b}(\mathbf{U})\,d\Omega = \sum_{k=1}^{NE} \mathbf{R}_k^{-1}\mathbf{H}(\hat{\mathbf{U}})_k l_k \qquad \int_{\Omega_i} \mathbf{S}_\tau(\mathbf{U})\,d\Omega = \sum_{k=1}^{NE} \mathbf{R}_k^{-1}\mathbf{T}(\hat{\mathbf{U}})_k l_k \tag{49}$$

where $\mathbf{H}(\hat{\mathbf{U}})_k$ and $\mathbf{T}(\hat{\mathbf{U}})_k$ are the integrated bed pressure [26] and basal resistance [27] throughout the kth cell edge, expressed in the local framework, being

$$\mathbf{H}(\hat{\mathbf{U}})_k = (0,\ -grh\ \Delta z_b,\ 0,\ 0,\ 0)_k^T \tag{50}$$

$$\mathbf{T}(\hat{\mathbf{U}})_k = \left(0, -\frac{\tau_b}{\rho_w}\left(n_{ux}\Delta x + n_{uy}\Delta y\right), 0, 0, 0\right)_k^T \tag{51}$$

Using (46) and (49), the augmented flux at the kth edge can be defined as

$$\mathcal{F}(\hat{\mathbf{U}})_k = \left(\mathbf{F}(\hat{\mathbf{U}}) - \mathbf{H}(\hat{\mathbf{U}}) - \mathbf{T}(\hat{\mathbf{U}})\right)_k \tag{52}$$

The net exchange flux term $\mathbf{E_b}(\mathbf{U})$ is discretized in space as

$$\int_{\Omega_i} \mathbf{E_b}(\mathbf{U})\,d\Omega \approx A_i\,\mathbf{E_b}(\mathbf{U}_i) \tag{53}$$

and, assuming a piecewise uniform representation of the variables at the i cell for the time $t = t^n$ and using explicit temporal integration for the mass and momentum source terms, the updating formulation for the conservative variables $\mathbf{U}$ is expressed as

$$\mathbf{U}_i^{n+1} = \mathbf{U}_i^n - \frac{\Delta t}{A_i}\sum_{k=1}^{\mathrm{NE}} \mathbf{R}_k^{-1}\mathcal{F}\left(\hat{\mathbf{U}}_i^n, \hat{\mathbf{U}}_j^n\right)_k^{\downarrow} l_k + \Delta t\,\mathbf{E_b}\left(\mathbf{U}_i^n\right) \tag{54}$$

being $\Delta t = t^{n+1} - t^n$ the time step, A_i the discrete cell area and $\mathcal{F}\left(\hat{\mathbf{U}}_i^n, \hat{\mathbf{U}}_j^n\right)_k^{\downarrow}$ the numerical flux through the kth cell edge computed as a function of the local conservative variables at the neighboring cells i and j. The numerical fluxes are here upwind computed using a fully-coupled Roe-type Riemann solver (RS) adapted to compressible two-phase shallow flows. Note that the flow density and depth remain coupled in both the conservative variables and fluxes on the left hand side of the equations, improving the robustness and accuracy of the solution when large density gradients appear in the flow [28]. The RS formulation is based on the augmented Roe strategy [25], ensuring the well-balance in steady states and the correct treatment of wet-dry fronts without requiring additional time step restrictions. A detailled explanation of this fully-coupled RS can be found in Aranda et al. [29].

3.2 High-performance-computing in GPU's

Graphics Processing Units (GPU's), were developed to control and manage the graphic operations for video games but currently they have become a powerful tool for solving engineering problems. The main advantage resides in the high number of computation nodes available for workload distribution, leading to higher speed-ups without an increment of investment in large facilities. The main drawback is that GPU-kernels usually require an intensive programming effort compared with multi-CPU strategies, such as OpenMP or other shared-memory strategies. There exist several platforms for High-Performance-Computing (HPC). NVIDIA developed the CUDA (Compute Unified Device Architecture), a computing platform and programming environment which incorporates directives for massive parallel operations. To ensure the applicability of the model to realistic large-scale mud/debris flows, the updating eq. (54) is solved using a GPU-based algorithm implemented in the NVIDIA CUDA/C++ framework.

The preprocess step and CPU-GPU memory transfer are implemented to run on one CPU core, whereas the time loop computation is accelerated using GPU. However, some tasks inside the time loop are controlled yet by the CPU, such as the time advance control, the boundary conditions application and the output data dump. Therefore, it is necessary to transfer information from/to the GPU at each time-step.

While the computational effort required for the time and boundaries transference is considerably smaller than that of each kernel function, in order to dump the intermediate output information, all the variables in the domain must be transferred from GPU device to CPU host.

The CUDA toolkit allows that all the processed elements can be distributed by threads and blocks of threads. Each thread uses its own thread index to identify the element to be processed, launching several execution threads at the same time (parallel computation). As computing GPU devices are well designed to work efficiently with ordered information, the variables needed for computation are stored in the GPU memory as structures of arrays (SoA), improving the spatial locality for memory accesses. Only the kernel functions, which require a higher computational effort, have been implemented to run on the GPU device. Some tasks in the GPU kernel are optimized using the CUBLAS library included in CUDA. The memory transfer between the CPU host and the GPU device has been reduced as much as possible for each time step.

4. Application to the mine tailings dam failure in Brumadinho (Brasil)

In this section, the proposed methodology is applied to the catastrophic large-scale mud flow occurred Brumadinho (Minas Gerais, Brazil, 2019). The sudden failure of a mine containing dike resul in an extremely violent tailings flow which traveled downstream more than 10 km and reached the Paraopeba River. This disaster caused more than 260 deaths and important economic and environmental losses. The dam contained almost $12 \cdot 10^6 m^3$ mining waste tailings with a height of $70 - 80m$ and covering an area of $4.13 \cdot 10^5 m^2$. **Figure 5** shows an aerial image of the mine site after

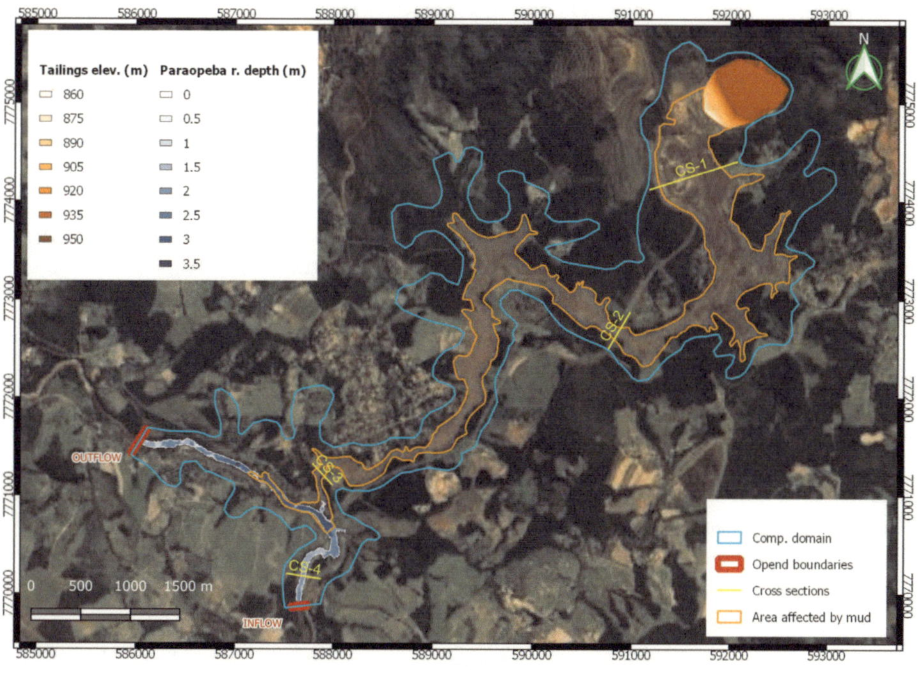

Figure 5.
Aerial image of the area affected by the mud.

Dam capacity		$12 \cdot 10^6$ m³		
Dam area		$4.13 \cdot 10^5$ m²		
Solid concent. ϕ_0		50%		
Specific weight		$22 - 26$ kN/m³		
Size distr.	Sand		Fines	
Rel. content	38%		62%	
Heavy metals	Fe	Al	Mn	Ti
Weight conc.	264.9 mg/g	10.8 mg/g	4.8 mg/g	0.5 mg/g

Table 1.
Brumadinho's dam and tailings features.

the dam collapse. The thalweg elevation along the area covered by the mud varies between $860 m.o.s.l$ at the dam-toe and $720\ m.o.s.l$ at the Paraopeba River, with an averaged longitudinal bed slope $S_0 = 0.0165\ m/m$. The area affected by the mud was $3.3 \cdot 10^6 m^2$.

The dike material and the tailings rapidly became a heavy liquid that flowed downstream at a high speed. Tailings were composed by a mixture of water, sediments and heavy metals, mainly iron, aluminum, manganese and titanium [30]. The size distribution consisted basically of a mineral sand fraction (38%) and a fines fraction (62%), accounting for mineral silt-clay and metals particles. The water content before the failure was estimated around 50% by volume with a specific weight of $22 - 26$ kN/m³ (**Table 1**).

The spatial domain is discretized using a unstructured triangular mesh with $6 \cdot 10^5$ cells approximately and four control cross-sections are placed downstream of the dam (**Figure 5**), corresponding to (CS-1) the mine stockpile area, (CS-2) the railway bridge, (CS-3) the national road and (CS-4) the gauge station in the river. Furthermore, the base regime water depth and velocities in the Paraopeba River before to the mud arrival is assessed by a previous simulation. The initial tailings depth at the dam is estimated by comparing the terrain elevation before and after the dam failure using 1×1 m DTM's. Six different solid phases are set for characterizing the fully saturated tailings material, including mineral sand, mineral silt, iron (Fe), aluminum (Al), manganese (Mn) and titanium (Ti), with initial bulk volumetric concentration $\phi_0 = 0.5$ and normalized density $r \approx 2.25$. The deposition porosity ξ for the bed layer is estimated using the Wu relation [8] and the hiding-exposure effects on the critical Shields stress for the incipient motion of each solid phase $\theta_{c,p}$ are estimated using the Egiazaroff formula [31]. Also, the solid transport capacity of the flow q_s^* is computed using the Wu relation [8]. The simulated time was 3 h from the dam collapse.

The mining tailings in the dam showed a low plasticity and high values of pore-fluid pressure [30] before the collapse. Assuming the turbulent behavior of the flow, the behavior index is set $m = 2$, setting a quadratic relation in (37) and allowing to calculate the apparent consistency parameter μ_P as a function of the Manning's roughness coefficient of the terrain ($n_b = 0.065 sm^{-1/3}$). The basal stability angle is estimated in $\delta_f = 5^\circ$. The basal pore pressure is estimated as $\mathcal{P}_b = (1 + \mathcal{E}_b)\rho_w g h$, being $\mathcal{E}_b$ a basal pressure factor which tends to $\mathcal{E}_b \to \infty$ for totally fluidized materials. Therefore, the total frictional-turbulent basal shear stress is estimated as

$$\frac{\tau_b}{\rho_w} = (r - 1 - \mathcal{E}_b)gh + rg\frac{n_b^2}{h^{1/3}}(u^2 + v^2) \tag{55}$$

where the first term on the right hand side accounts for the frictional yield stress whereas the second term denotes the turbulent shear component.

In order to assess the influence of the basal pressure in the numerical results, the entrainment term $\mathbf{E_b}(\mathbf{U})$ in (42) is neglected and the value of $\mathcal{E}_b$ is varied from $\mathcal{E}_b \to \infty$ (pure Newtonian turbulent behavior) to $\mathcal{E}_b = 0.9$ (medium frictional yield strength). **Figure 6** shows the temporal evolution of the wave-front location and the total flow rate at the control section CS-2. Note that, for each value of $\mathcal{E}_b$, the wave front loses velocity progressively as it moves downstream. As $\mathcal{E}_b$ decreases, the frictional shear stress is enhanced and the mobility of the flow decreases considerably. That it is clearly shown in the marked reduction of the flow rate at the control section CS-2 as the pore pressure factor is increased. Setting $\mathcal{E}_b = 1.05$ allows to predict the observed arrival time ($32 - 47$ min) of the tailings wave to the Paraopeba river at 8.5 km downstream the dam (gray rectangle in **Figure 6**).

Figure 7 shows the mud depth at $t = 0$ min , $t = 5$ min , $t = 15$ min and $t = 45$ min after the dam collapse for $\mathcal{E}_b = 1.05$. The numerical results show that practically the whole initial tailing volume flows out of the dam in the first 5 min after the dam collapse, as it was observed in the available videos. The mud wave moves downstream with a computed height larger than 20 m in some zones during the first minutes. After this initial stage, the numerical results show that the dambreak wave is still higher than 10 m when the wave-front reaches the railway bridge (CS-2). During the real event, the mud wave impact caused the collapse of the railway bridge structure. At $t = 45$ min , the mud wave-front has reached the Paraopeba River and the flow is practically stopped.

Furthermore, in order to assess the influence of the solid material entrainment from the erodible bed to the flow layer, the pore-pressure factor is set to $\mathcal{E}_b = 1.05$ but the entrainment term $\mathbf{E_b}(\mathbf{U})$ in (42) is now considered. The relation α_E/α_D, which controls the balance between the erosion component E_b and the deposition component D_b of the vertical bed-flow exchange flux, is varied to $\alpha_E/\alpha_D = 1\%$, $\alpha_E/\alpha_D = 5\%$, $\alpha_E/\alpha_D = 10\%$ and $\alpha_E/\alpha_D = 50\%$. **Figure 8** depicts the predicted temporal evolution of both the mud wave-front location and the total volume of the fluidized material in the domain as the entrainment parameter is varied. The higher the relation α_E/α_D, the larger the fluidized mass involved in the flow as the dambreak wave progresses. Hence, the entrainment of bed material leads to a greater flow mobility, causing the wave-front to reach the Paraopeba River faster as the entrainment parameter α_E/α_D

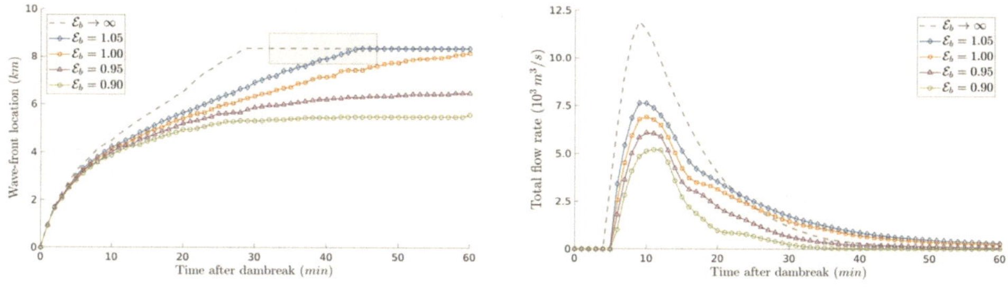

Figure 6.
Temporal evolution of (left) the wave-front location and (right) the total flow rate at the control section CS-2.

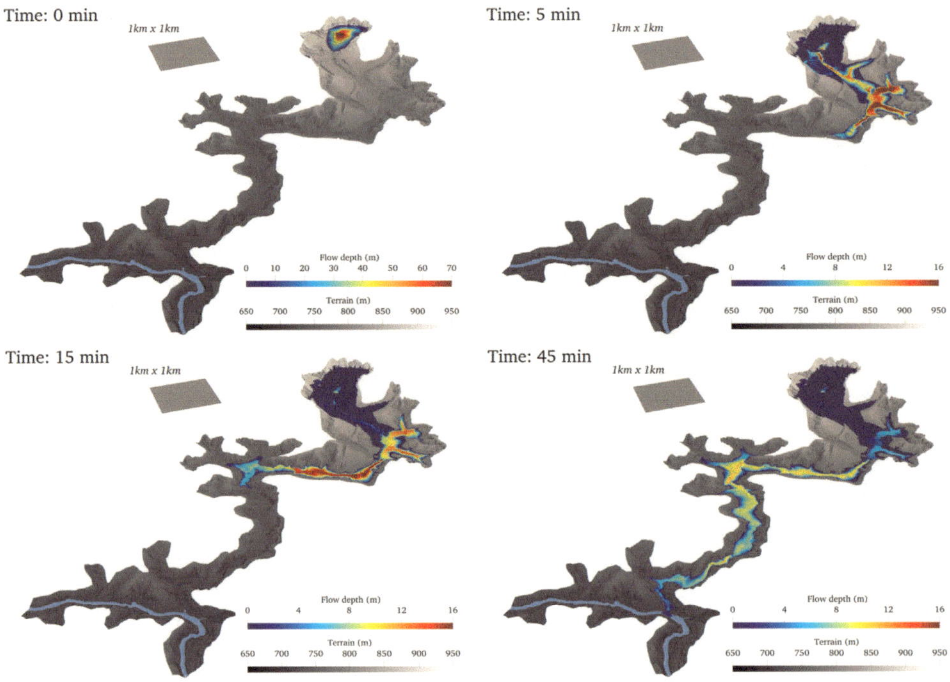

Figure 7.
Mud depth at $t = 0$ min, $t = 5$ min, $t = 15$ min *and* $t = 45$ min *after the dam collapse.*

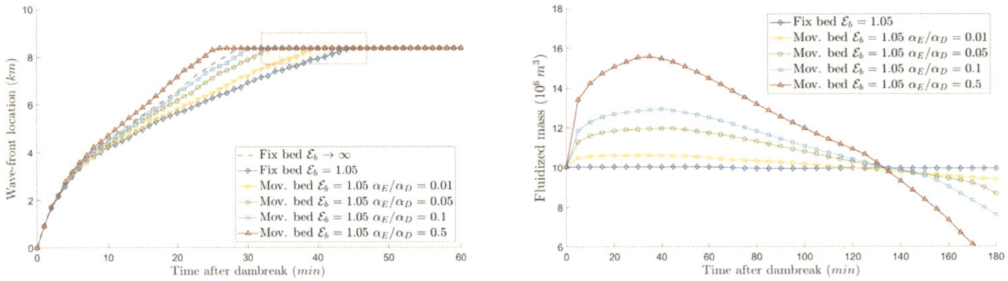

Figure 8.
Temporal evolution of (left) the wave-front location and (right) the total fluidized volume as the relation α_E/α_D *is increased.*

increases. Even, for $\alpha_E/\alpha_D = 50\%$, the mobilized mass becomes more than 150% of the initial mass and the wave-front reaches the river faster than the fixed-bed pure-turbulent case (dashed gray line).

Furthermore, the entrainment of bed materials also leads to the segregation of the solid phases along the flow. As α_E/α_D is increased, horizontal gradients in the concentration of the different solid classes composing the mixture tend to appear (**Figure 9**). These gradients depends on the temporal evolution of the class-specific erosion and deposition vertical fluxes, E_p and D_p respectively, along the tailings flow.

The increment of the flow mobility caused by the bed material entrainment is clearly shown in the temporal evolution of the flow rate at the control sections CS-2 and CS-3 (**Figure 10**). At CS-2 the increment of the entrainment relation α_E/α_D is mainly appreciated in the increment of the flow rate peak from 7500 m^3/s for the

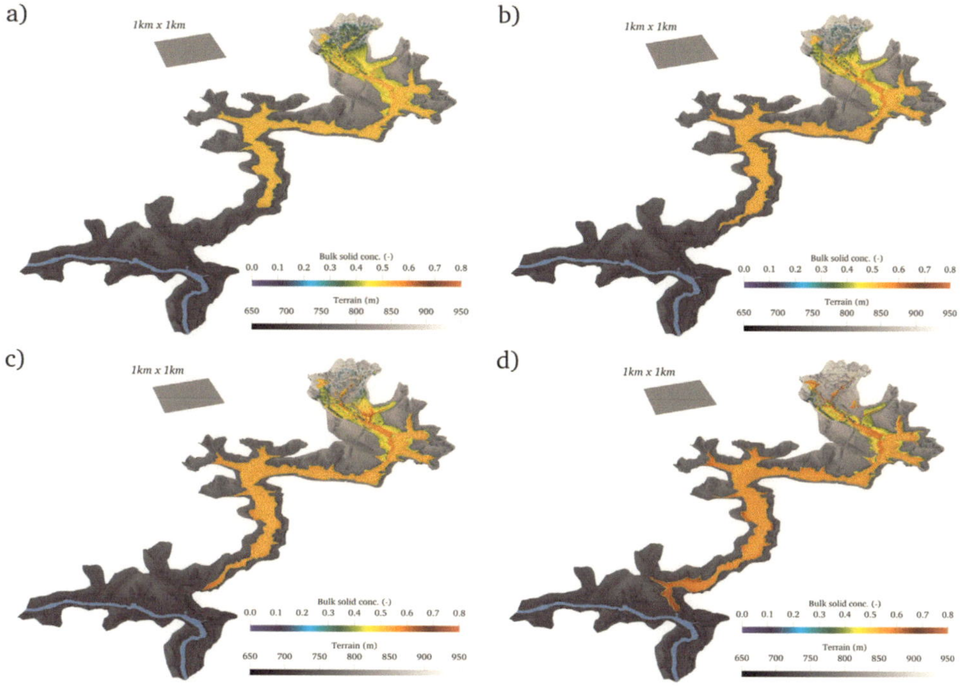

Figure 9.
Bulk solid concentration in the flow at $t = 25$ min *after the dam collapse for (a)* $\alpha_E/\alpha_D = 1\%$*, (b)* $\alpha_E/\alpha_D = 5\%$*, (c)* $\alpha_E/\alpha_D = 10\%$ *and (d)* $\alpha_E/\alpha_D = 50\%$*.*

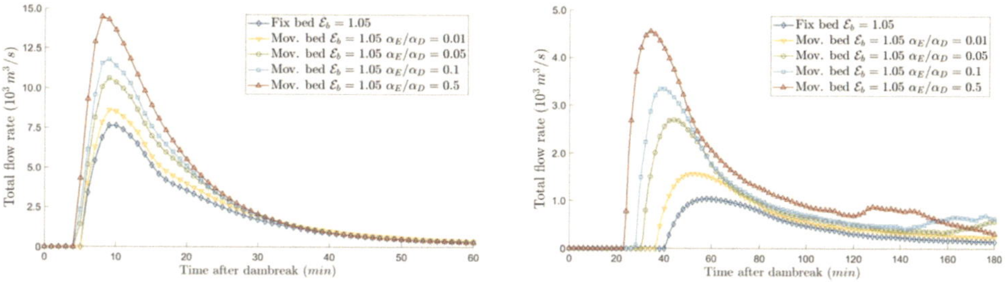

Figure 10.
Temporal evolution of the flow rate at (left) the control sections CS-2 and (right) the control section CS-3 as α_E/α_D *is increased.*

fixed-bed simulation to almost 15,000 m^3/s for the movable bed case with $\alpha_E/\alpha_D = 50\%$. Furthermore, at CS-3, the entrainment of bed material does not only increase the flow rate peak respect to the fixed-bed assumption, but also leads to a noticeable decrease in the arrival time of the mud wave from 40 min to 25 min approximately.

Finally, the computational efficiency of the GPU-accelerated kernel is shown in **Table 2** for different GPU devices. The performance obtained with the GPU-based code is compared with the single-core CPU simulation time. Note that the speed-up obtained respect to the single-core performance has increased progressively as the devices have been developed in the last 5 years. Currently, the speed-up obtained with

	Computational time	Speed-up
CPU Intel i7-10700F	59.43 h	—
Nvidia Tesla K40c	1.79 h	× 33
Nvidia GTX 1080 Ti	58.2 min	× 61
Nvidia Tesla V100	26.5 min	× 135
Nvidia A100 40GB	12.8 min	× 279

Table 2.
Computational times with GPU-based and CPU-based algorithms.

the GPU accelerated kernel is about 280 (Nvidia A100), more than 8 times the Nvidia Tesla K40c performance. That means that for achieving the GPU-parallelized code performance with a CPU-based algorithm, a cluster with at least 280 CPU cores is required (see speed-up in **Table 2**).

5. Conclusions

The numerical modeling of geophysical surface flows involving sediment transport must be addressed using a comprehensive strategy: beginning with the derivation of proper shallow-type mathematical models, following by the development of robust and accurate numerical schemes within the Finite Volume (FV) framework and ending with the implementation of efficient HPC algorithms. This integrated approach is required to release Efficient Simulation Tools (EST) for environmental processes involving sediment transport with realistic temporal and spatial scales.

The derivation of the generalized 2D system of depth-averaged conservation laws for environmental surface flows of water-sediment mixtures over movable bed conditions is a fundamental step to understand the physical consequences of the mathematical shallow-type simplification. From the mathematical modeling approach, the fluidized material in motion is contained in a flow layer consisting of a mixture of water and multiple solid phases. This flow layer usually moves rapidly downstream steep channels and involves complex topography. The liquid-solid material can be exchanged throughout the bottom interface with the underlying static bed layer, hence involving also a transient bottom boundary for the flow layer. These features lead to an increasing complexity for the mathematical simplification of sediment transport surface flows.

The 2D shallow-type system of equations for variable-density multi-grain water-sediment flows can be solved using a Finite Volume (FV) methods, supplemented preferable with upwind augmented Riemann solvers which provide a robust and accurate computation of the intercell fluxes even involving highly transient density interfaces and ensure the well-balanced character of the solution in quiescent and steady states. The usage of GPU-accelerated kernels for sequential computation of the FV solution is an efficient and low cost alternative to the traditional multi-CPU strategies. The GPU solution must be designed taking into account the fact that the GPU is an independent device with its own RAM memory. This means that the memory transfer between the conventional RAM memory and the GPU device memory plays a key role in the performance of GPU-accelerated software.

GPU-accelerated codes are mandatory for large-scale sediment-laden flow models since they allows the calibration of the multiple processes involved in the movement of the fluidized water-sediment material without spending weeks or even months waiting results. Probably, the most challenging and unknown process involved in sediment-laden surface flows is the development of pore-fluid pressure, hence its estimation requires careful calibration procedure. This pore pressure affects the frictional shear stress between solid particles, reducing the basal resistance and affecting the flow mobility. Furthermore, special attention has to be paid to the calibration of the entrainment/deposition parameters. The entrainment of material from the underlying movable bed to the mixture layer increases the mass and momentum of the fluidized material in movement, enlarging the mobility of the flow and reducing the arrival time of the mud waves.

Acknowledgements

This work is part of the research project PGC2018-094341-B-I00 Fast and accurate tools for the simulation and control of environmental flows (FACTOOL) funded by the Ministry of Science and Innovation and FEDER.

Author details

Sergio Martínez-Aranda*† and Pilar García-Navarro†
Fluid Dynamic Technologies (i3A), University of Zaragoza, Zaragoza, Spain

*Address all correspondence to: sermar@unizar.es

† These authors contributed equally.

References

[1] Iverson RM. The physics of debris flows. Reviews of Geophysics. 1997; **35**(3):245-296

[2] Pierson TC. Hyperconcentrated flow—Transitional process between water flow and debris flow. In: Debris-flow Hazards and Related Phenomena. Berlin, Germany: Springer Berlin Heidelberg; 2005

[3] Hungr O, Evans SG, Bovis MJ, Hutchinson JN. A review of the classification of landslides of the flow type. Environmental and Engineering Geoscience. 2001;**7**(3):221-238

[4] Rickenmann D, Weber D, Stepanov B. Erosion by debris flows in field and laboratory experiments. In: 3rd International Conference on Debris-Flow Hazards: Mitigation, Mechanics, Prediction and Assessment. Rotterdam: Millpress; 2003. pp. 883-894

[5] Iverson RM, Reid ME, Logan M, LaHusen RG, Godt JW, Griswold JP. Positive feedback and momentum growth during debris-flow entrainment of wet bed sediment. Nature Geoscience. 2011;**4**:116-121

[6] Berger C, McArdell BW, Schlunegger F. Direct measurement of channel erosion by debris flows, illgraben, Switzerland. Journal of Geophysical Research: Earth Surface. 2011;**116**(F1):93-104

[7] McCoy SW, Kean JW, Coe JA, Tucker GE, Staley DM, Wasklewicz TA. Sediment entrainment by debris flows: In situ measurements from the headwaters of a steep catchment. Journal of Geophysical Research: Earth Surface. 2012;**117**:F03016

[8] Wu W. Computational River Dynamics. NetLibrary, Inc. London: CRC Press; 2007

[9] Murillo J, García-Navarro P. Wave riemann description of friction terms in unsteady shallow flows: Application to water and mud/debris floods. Journal of Computational Physics. 2012;**231**: 1963-2001

[10] Juez C, Murillo J, García-Navarro P. 2d simulation of granular flow over irregular steep slopes using global and local coordinates. Journal of Computational Physics. 2013;**255**: 166-204

[11] Gualtieri C, Ianniruberto M, Filizola N. On the mixing of rivers with a difference in density: The case of the negro/solimões confluence, Brazil. Journal of Hydrology. 2019;**578**: 124029

[12] Cunge JA, Holly FM, Verwey A. Practical aspects of computational river hydraulics. In: Monographs and Surveys in Water Resources Engineering. London: Pitman Advanced Publishing Program; 1980

[13] Dewals B, Rulot F, Erpicum S, Archambeau P, Pirotton M. Sediment Transport, Chapter Advanced Topics in Sediment Transport Modelling: Non-alluvial Beds and Hyperconcentrated Flows. Rijeka, Croatia: IntechOpen; 2011

[14] Lacasta A, Juez C, Murillo J, García-Navarro P. An efficient solution for hazardous geophysical flows simulation using GPUs. Computers & Geosciences. 2015;**78**:63-72

[15] Ming X, Liang Q, Xia X, Li D, Fowler HJ. Real-time flood forecasting based on a high-performance 2d hydrodynamic model and numerical weather predictions. Water Resources Research. 2020;**56**:e2019WR025583

[16] Juez C, Lacasta A, Murillo J, García-Navarro P. An efficient GPU implementation for a faster simulation of unsteady bed-load transport. Journal of Hydraulic Research. 2016;**54**(3):275-288

[17] de la Asuncion M, Castro MJ. Simulation of tsunamis generated by landslides using adaptive mesh refinement on gpu. Journal of Computational Physics. 2017;**345**:91-110

[18] Pastor M, Blanc T, Haddad B, Drempetic V, Sanchez-Morles M, Dutto P, et al. Depth averaged models for fast landslide propagation: Mathematical, rheological and numerical aspects. Archives of Computational Methods in Engineering. 2015;**22**: 67-104

[19] Jakob M, Hungr O. Debris-Flow Hazards and Related Phenomena. Springer Praxis Books. Springer Berlin Heidelberg; 2005

[20] Iverson RM, Vallance JW. New views of granular mass flows. Geology. 2001;**29**(2):115-118

[21] Berti M, Simoni A. Experimental evidences and numerical modeling of debris flow initiated by channel runoff. Landslides. 2005

[22] McArdell BW, Bartelt P, Kowalski J. Field observations of basal forces and fluid pore pressure in a debris flow. Geophysical Research Letters. 2007; **34**(7):L07406

[23] Richardson JF, Zaki WN. Sedimentation and fluidisation: Part I. Chemical Engineering Research and Design. 1997;**75**:82-100

[24] Godlewski E, Raviart P-A. Numerical Approximation of Hyperbolic Systems of Conservation Laws. New York: Springer-Verlag; 1996

[25] Murillo J, Navas-Montilla A. A comprehensive explanation and exercise of the source terms in hyperbolic systems using Roe type solutions. Application to the 1D-2D shallow water equations. Advances in Water Resources. 2016;**98**:70-96

[26] Murillo J, García-Navarro P. Energy balance numerical schemes for shallow water equations with discontinuous topography. Journal of Computational Physics. 2012;**236**:119-142

[27] Martínez-Aranda S, Murillo J, Morales-Hernández M, García-Navarro P. Novel discretization strategies for the 2D non-Newtonian resistance term in geophysical shallow flows. Engineering Geology. 2022;**302**:106625

[28] Martínez-Aranda S, Murillo J, García-Navarro P. A robust two-dimensional model for highly sediment-laden unsteady flows of variable density over movable beds. Journal of Hydroinformatics. 2020;**22**(5):1138-1160

[29] Martínez-Aranda S, Murillo J, García-Navarro P. A GPU-accelerated efficient simulation tool (EST) for 2D variable-density mud/debris flows over non-uniform erodible beds. Engineering Geology. 2021;**296**:106462

[30] Robertson PK, de Melo L, Williams DJ, Ward Wilson G. Report of the Expert Panel on the Technical Causes of the Failure of Feijão Dam I. Technical Report. Brasil: Vale S.A.; 2019

[31] Egiazaroff IV. Calculation of nonuniform sediment concentrations. Proceedings of the American Society of Civil Engineers. 1965;**91**:225-247